Arithmetic in the Thought of Gerbert of Aurillac

PHILOSOPHY AND CULTURAL STUDIES REVISITED / HISTORISCH-GENETISCHE STUDIEN ZUR PHILOSOPHIE UND KULTURGESCHICHTE

Edited by
Seweryn Blandzi

Advisory Board

Manfred Frank (University of Tübingen)
Kamila Najdek (University of Warsaw)
Marek Otisk (University of Ostrava)
Wojciech Starzyński (Polish Academy of Sciences)

VOLUME 10

Marek Otisk

Arithmetic in the Thought of Gerbert of Aurillac

Bibliographic Information published by the Deutsche Nationalbibliothek
The Deutsche Nationalbibliothek lists this publication in the Deutsche Nationalbibliografie; detailed bibliographic data is available online at http://dnb.d-nb.de.

Library of Congress Cataloging-in-Publication Data
A CIP catalog record for this book has been applied for at the Library of Congress.

This publication was financially supported by the Faculty of Arts at the University of Ostrava.

Cover illustration: Photo by Marek Otisk.

ISSN 2510-5353
ISBN 978-3-631-85816-5 (Print)
E-ISBN 978-3-631-87071-6 (E-PDF)
E-ISBN 978-3-631-87089-1 (EPUB)
DOI 10.3726/b19269

© Peter Lang GmbH
Internationaler Verlag der Wissenschaften
Berlin 2022
All rights reserved.

Peter Lang – Berlin · Bern · Bruxelles · New York · Oxford · Warszawa · Wien

All parts of this publication are protected by copyright. Any utilization outside the strict limits of the copyright law, without the permission of the publisher, is forbidden and liable to prosecution. This applies in particular to reproductions, translations, microfilming, and storage and processing in electronic retrieval systems.

This publication has been peer reviewed.

www.peterlang.com

To my parents

Contents

PREFACE 9

INTRODUCTION: ARITHMETIC AS THE SCIENCE OF NUMBERS 15

I. THEORETICAL ARITHMETIC 33

    1. The Nature of Numbers: *Saltus Gerberti* (Letter to Constantine) 34

    2. Figurative Numbers and Geometry (Letter to Adelbold) 55

    3. Music and Harmony (Two Letters to Constantine) 69

    4. Astronomy and Timekeeping (Letter *De sphaera*, Richer's *Historia*, and the Horological letter to Adam) 81

II. PRACTICAL ARITHMETIC 129

    1. *Tabula abaci*, Decimal Positional Notation and *Ghubar* Numbers ... 129

    2. *Regulae multiplicationis* 156

    3. *Regulae divisionis* 170

CONCLUSION 185

BIBLIOGRAPHY 189

    Abbreviations 189

    Primary Sources 189

    Secondary Sources 198

INDICES 217

List of Tables 219

List of Figures 221

Index locorum 223

Index of Personal Names (before 1700) 233

Index of Personal Names (after 1700) 237

PREFACE

Since the 19th century, multiple interpretative strategies have portrayed the long period of the European Latin Middle Ages as a consecutive succession of renaissances.[1] Despite the vagueness of such a designation, and even although it is difficult to summarise the vast diversity of historical events it denotes, it has become – pace all justified criticism – the dominant interpretative model applied to medieval intellectual and cultural history.[2] Various terms have emerged over time, including the Carolingian Renaissance (*Karolingische Renaissance*),[3] and its phases,[4] which were supposedly preceded by the renaissances of the 7th century in, for example, Visigoth Hispania or the Irish monasteries and monasteries on the British Isles, which directly influenced the Carolingian Renaissance.[5] If we

---

1 One of the original uses of the term renaissance in relation to the intellectual and cultural development of the Middle Ages can be found in texts by Jean-Jacques Ampère – see Ampère, J.-J., "Histoire littéraire de la France avant le douzième siècle." *Revue des Deux Mondes* 5 (1936), pp. 34–35; or idem, *Histoire littéraire de la France avant le douzième siècle*. Vol. 3. Paris: Hachette, 1840, p. 457. On the issue, see, e.g., Holmes, U. T., "The Idea of a Twelfth-Century Renaissance." *Speculum* 26/4 (1951), pp. 643–644.
2 See, e.g., Wood, I., "Culture." In: McKitterick, R. (ed.), *The Early Middle Ages. Europe 400–1000*. Oxford: Oxford University Press, 2001, p. 168; Ferguson, W. K., *The Renaissance in Historical Thought. Five Centuries of Interpretation*. Boston – New York et al.: Houghton Mifflin Company, 1948, especially pp. 329–385; Le Goff, J., *Les Intellectuels au Moyen Âge*. Paris: Éditions du Seuil, 1957, for example pp. 10–14; and so on.
3 Cf. Trompf, G. W., "The Concept of the Carolingian Renaissance." *Journal of the History of Ideas* 34/1 (1973), pp. 3–26; Patzelt, E., *Die Karolingische Renaissance. Beiträge zur Geschichte der Kultur des Frühen Mittelalters*. Wien: Österreichischer Schulbücherverlag, 1924; Contreni, J., "The Carolingian Renaissance: Education and Literary Culture." In: McKitterick, R. (ed.), *The New Cambridge Medieval History*. Vol. 2: *c. 700–c. 900*. Cambridge: Cambridge University Press, 2015, pp. 709–757; Marenbon, J., *Early Medieval Philosophy*. London – New York: Routledge, 1988, pp. 46–79; Nelson, J. L., "On the Limits of the Carolingian Renaissance." *Studies in Church History* 14 (1977): *Renaissance and Renewal in Christian History*, pp. 51–69; etc.
4 Cf., for example, Riché, P. – Verger, J., *Des nains sur des épaules de géants. Maîtres et élèves au Moyen Âge*. Paris: Tallandier, 2006, pp. 31–74.
5 Cf., for instance, Mülke, M., "'Isidorische Renaissance' – oder: Über die Anbahnung einer Wiedergeburt." *Antiquité Tardive* 23 (2015), pp. 95–107; Polara, G., "Secolo VII." In: Leonardi, C. (ed.), *Letteratura latina medievale (secoli VI–XV). Un manuale*. Firenze: SISMEL – Edizioni del Galluzzo, 2002, pp. 17–40; O'Connor, R., "Irish narrative

divide the Carolingian Renaissance into several stages, we may distinguish the era of Charlemagne (the last third of the 8th century and the start of the 9th century), followed by the period after Charlemagne (especially the reign of Louis the Pious and Charles the Bald in the 9th century), and then the so-called Renaissance of the 10th century or the Ottonian Renaissance (*Ottonische Renaissance*),[6] which overlaps with the so-called Big Restoration after the year 1000,[7] and encompasses mainly the second half of the 10th century and the first quarter of the 11th century.[8] This is succeeded by the Renaissance of the 11th century,[9] in which the foundations were laid for the (most famous) medieval Renaissance of the 12th century and for the paradigmatic constants that remained typical of the scholastic era of the Middle Ages.[10]

---

    literature and the Classical tradition." In: O'Connor, R. (ed.), *Classical Literature and Learning in Medieval Irish Narrative*. Cambridge: D. S. Brewer, 2014, pp. 1–2; Herren, M., "Classical and Secular Learning among the Irish before the Carolingian Renaissance." *Florilegium* 3 (1981), pp. 118–157; etc.

6 Cf. Naumann, H., *Karolingische und Ottonische Renaissance*. Frankfurt a. M.: Englert und Schlosse, 1926; Rückert, O., *Ottonische Renaissance. Ausgewählte Stücke aus Widukind von Corvey, Ruotger, Liudprand von Cremona, Hrotsvit von Gandersheim, Ekkehard IV. von St. Gallen*. Leipzig: Teubner, 1926; D'Onofrio, G., "Introduzione." In: D'Onofrio, G. (ed.), *Excerpta Isagogarum et Categoriarum*. CCCM 120. Turnhout: Brepols, 1995, pp. LXXXVII–CVI; McKitterick, R., "Ottonian intellectual culture in the tenth century and the role of Theophano." In: Davids, A. (ed.), *The Empress Theophano: Byzantium and the West at the turn of the first millenium*. Cambridge: Cambridge University Press, 1995, pp 169–193; etc.

7 Cf., for example, Frassetto, M. (ed.), *The Year 1000: Religious nad Social Response to the Turning of the First Millennium*. New York: Palgrave Macmillan, 2002, p. 134; Landes, R., "The Fear of an Apocalyptic Year 1000: Augustinian Historiography, Medieval and Modern." *Speculum* 75/1 (2000), p. 134; idem, "Rudolfus Glaber and the Dawn of the New Millenium: Eschatology, Historiography, and the Year 1000." *Revue Mabillon. Revue Internationale d'Histoire et de Littérature Religieuses* 7 (1996), pp. 57–77; etc.

8 See, e.g., Truitt, E. R., "Celestial Divination and Arabic Science in Twelfth-Century England: The History of Gerbert of Aurillac's Talking Head." *Journal of the History of Ideas* 73/2 (2012), p. 214; or Catalani, L., "«Omnia Numerorum Videntur Ratione Formata». A 'Computable World' Theory in Early Medieval Philosophy." In: Gadducci, F. – Tavosanis, M. (eds.), *History and Philosophy of Computing*. Berlin – Cham: Springer, 2016, pp. 131–140.

9 Cf., for instance, Vaughn, S. N. – Rubenstein, J. (eds.), *Teaching and Learning in Northern Europe, 1000–1200*. Turnhout: Brepols, 2006; Williams, J. R., "The Cathedral School of Rheims in the Eleventh Century." *Speculum* 29/4 (1954), pp. 661–677.

10 See the widely recognized revolutionary work: Haskins, Ch. H., *The Renaissance of the Twelfth Century*. Cambridge: Harvard University Press, 1927. Cf. also Benson,

In these interpretative models, the Early Middle Ages, formerly branded a dark era of European intellectual and cultural history,[11] became the succession of gradual milestones that essentially evidence the continual development of the ancient, as well as Byzantine and Arabic, intellectual heritage in the Latin West. Although it is possible to identify different forms of overlapping cultural and intellectual influences with varying degrees of intensity throughout the entire Middle Ages, this book focuses on the period labelled the "Ottonian Renaissance" or "the Renaissance of the second half of the 10th century," when we have evidence not only of the development of ancient intellectual heritage in the Latin West but also of numerous contacts with the Byzantine (e.g., with the imperial court) and Arabic (especially the Iberian peninsula) intellectual environments. This historical era of the European West owes its name to the Liudolfinger imperial dynasty, originally Saxon dukes and, later, kings of East Francia, who were crowned as Roman Emperors; the first emperor of this dynasty was Otto I (crowned Holy Roman Emperor by Pope John XII in 962) who was succeeded by his son Otto II (crowned by Pope John XIII in 967), his grandson Otto III (crowned by Pope Gregory V in 996), and his great-grandson Henry II, known as Saint Henry the Exuberant (crowned by Pope Benedict VIII in 1014).[12]

Among the most well-known contemporary figures with an intellectual (but also ecclesiastical, and often even political) background at the end of the 10th century and the beginning of the 11th century, it is possible to mention Notker of Liège († 1008, bishop from 972 and prince-bishop from 980 in Liège),[13] Abbo of

---

R. L. – Constable, G. (eds.), *Renaissance and renewal in the twelfth century*. Toronto – Buffalo – London: University of Toronto Press, 1999; Novikoff, A., "The Renaissance of the Twelfth Century Before Haskins." *The Haskins Society Journal. Studies in Medieval History* 16 (2005), pp. 104–116; Holmes, "The Idea…," pp. 643–651; Melve, L., " 'The revolt of the medievalists'. Directions in recent research on the twelfth-century renaissance." *Journal of Medieval History* 32 (2006), pp. 231–252; Swanson, R. N., *The Twelfth-Century Renaissance*. Manchester: Manchester University Press, 1999; Novikoff, A. J., "Anselm, Dialogue, and the Rise of Scholastic Disputation." *Speculum* 86/2 (2011), pp. 387–418; and many others.

11 For a brief historical overview of this term in relation to the Middle Ages, see, e.g., Nelson, J. L., "The Dark Ages." *History Workshop Journal* 63/1 (2007), pp. 191–201.

12 For a historical overview, see, e.g., Baumann, H., *Die Ottonen*. Stuttgart: Kohlhammer, 1987; Schulmeyer-Ahl, K., *Der Anfang vom Ende der Ottonen. Konstitutionsbedingungen historiographischer Nachrichten in der Chronik Thietmars von Merseburg*. Berlin – New York: De Gruyter, 2009; Wangerin, L. E., *Kingship and Justice in the Ottonian Empire*. Ann Arbor: University of Michigan Press, 2019; etc.

13 For more details, see, e.g., Delville, J.-P. – Kupper, J.-L.– Laffineur-Crepin, M. (eds.), *Notger et Liège. L'an mil au cœur de l'Europe*. Liège: Éditions du Perron, 2008; Kurth,

Fleury († 1004, abbot from 988 in Fleury),[14] Gerbert of Aurillac († 1003, abbot of Bobbio from 982, illegitimate archbishop of Reims from 991, archbishop of Ravenna from 998, Pope Sylvester II from 999),[15] Fulbert of Chartres († 1028, bishop of Chartres from 1006),[16] or Hermann the Lame († 1054, a monk of Reichenau monastery from 1020).[17] For all the above, it is the case that their

---

G., *Notger de Liège et la civilisation au 10ᵉ siècle*. Vol. 1. Paris – Bruxelles – Liège: A. Picard, 1905; Lutz, C. E., *Schoolmasters of the Tenth Century*. Hamden: Archon Books, 1977, pp. 93–99.

[14] For more details, see, e.g., Dachowski, E., *First Among Abbots: The Career of Abbo of Fleury*. Washington: Catholic University of America Press, 2008; Riché, P., *Abbon de Fleury. Un moine savant et combatif (vers 950–1004)*. Turnout: Brepols, 2004; Engelen, E.-M., *Zeit, Zahl und Bild. Studien zur Verbindung von Philosophie und Wissenschaft bei Abbo von Fleury*. Berlin – New York: De Gruyter, 1993; Mostert, M., *The political theology of Abbo of Fleury: a study of the ideas about society and law of the tenth-century monastic reform movement*. Hilversum: Verloren, 1987; and many others.

[15] For more detailed information about the life, work, or so-called legend of Gerbert, see, e.g., Brown, N. M. *The Abacus and the Cross. The Story of the Pope Who Brought the Light of Science to the Dark Ages*. New York: Basic Books, 2010; Flusche, A. M., *The Life and Legend of Gerbert of Aurillac. The Organbuilder Who Became Pope Sylvester II*. Lewiston: Edwin Mellen Press, 2005; Riché, P., *Gerbert d'Aurillac: Le pape de l'an mil*. Paris: Fayard, 1987; Tosi, M. (ed.), *Gerberto – scienza, storia e mito. Atti del Gerberti Symposium*. Bobbio: A.S.B., 1985; Oldoni, M., "Gerberto e la sua Storia." *Studi Medievali* 18/2 (1977), pp. 629–704; idem, "'A fantasia dicitur fantasma' (Gerberto e la sua storia, II)." *Studi Medievali* 21/2 (1980), pp. 493–622; idem, "'A fantasia dicitur fantasma' (Gerberto e la sua storia, II) II." *Studi Medievali* 24/1 (1983), pp. 167–245; Darlington, O. G., "Gerbert, the Teacher." *The American Historical Review* 52/3 (1947), pp. 456–476; and many others.

[16] For further reading, cf. Genin, C., *Fulbert de Chartres (vers 970–1028): Une grande figure de l'Occident chrétien au temps de l'an Mil*. Chartres: Société Archéologique d'Eure-et-Loir, 2003; MacKinney, L. C., *Bishop Fulbert and Education at the School of Chartres*. Notre Dame: University of Notre Dame, 1957; Erlande-Brandenburg, A. – Cazeaux, M. (eds.), *Le temps de Fulbert; Fulbert et les écoles de Chartres. Actes de l'Université d'été du 8 au 10 juillet 1996*. Chartres: Société Archéologique d'Eure-et-Loir, 1996; Rouche, M. (ed.), *Fulbert de Chartres. Précurseur de l'Europe médiévale?*. Paris: Presses de l'Univeristé Paris-Sorbonne, 2008; and others.

[17] About his life and work, see, e.g., Borst, A., "Ein Forschungsbericht Hermanns des Lahmen." *Deutsches Archiv für Erforschung des Mittelalters* 40 (1984), pp. 379–477; Berschin, W. – Hellmann, M., *Hermann der Lahme. Gelehrter und Dichter (1013–1054)*. Heidelberg: Mattes, ²2005; Heinzer, F. – Zotz, T. (eds.), *Hermann der Lahme. Reichenauer Mönch und Universalgelehrter des 11. Jahrhunderts*. Stuttgart: Kohlhammer, 2016; and many others.

scholarly and philosophical activities were focused mainly on the seven liberal arts. Their attention was variously divided between the sciences of the *trivium* (primarily dialectics, but grammar and rhetoric were not neglected) and the sciences of the *quadrivium* (i.e., arithmetic, geometry, music, and astronomy). This commitment is apparent in the contemporaries (and, to a certain extent, rivals) Abbo of Fleury and Gerbert of Aurillac – both were, among other things, authors of texts on dialectics (see, for instance, Abbo's texts on categorical and hypothetical syllogism, although he also wrote about grammar etc.,[18] or Gerbert's work *De rationali et ratione uti*[19]) on mathematics, and astronomy (Abbo is the author of computistic treatises, a commentary on the *Calculus* of Victorius, and several astronomical texts,[20] while Gerbert wrote, for example, several "scientific" letters to his friend Constantine of Fleury and Micy, including the famous *Regula de abaco computi*, or the construction manual *De sphera*, and several other texts, predominantly dealing with topics related to the arts of the *quadrivium*).[21]

This book primarily focuses on Gerbert of Aurillac (sometimes known as Gerbert "of Reims," "of Bobbio," or "of Ravenna," and also known by his papal name Sylvester II). The book's content summarises my systematic long-term interest in the period of pre-scholastic Latin thought, especially regarding the relationship between dialectics, metaphysics, natural philosophy, and theology in the 10th and 11th centuries. The topical focus of this book is directed at the

---

18 Cf., for instance, Riché, *Abbon de Fleury…*, pp. 282, 285; or Van de Vyver, A., "Les œuvres inédites d'Abbon de Fleury." *Revue Bénédictine* 47 (1935), pp. 130–140.
19 Cf., e.g., Sigismondi, F., *Gerberto d'Aurillac, il trattato De Rationali et Ratione Uti e la Logica del X secolo*. Roma: Ateneo Pontificio Regina Apostolorum, 2007; or Frova, C., "Gerberto philosophus: il De rationali et ratione uti." In: Tosi (ed.), *Gerberto – scienza, storia e mito…*, pp. 351–377.
20 See, for example, Van de Vyver, "Les œuvres inédites…," pp. 140–158; idem, "Les plus anciennes Traductions latines médiévales (X^e–XIe siècles) de Traités d'Astronomie et d'Astrologie." *Osiris* 1 (1936), pp. 677–678; Thomson, R. B., "Two Astronomical Tractates of Abbo of Fleury." In: North, J. D. – Roche, J. J. (eds.), *The Light of Nature. Essays in the History and Philosophy of Science presented to A. C. Crombie*. Dordrecht – Boston – Lancaster: Martinus Nijhoff Publishers, 1985, pp. 113–133; idem, "Further Astronomical Material of Abbo of Fleury." *Mediaeval Studies* 50 (1988), pp. 671–673; Evans, G. R. – Peden, A. M., "Natural Science and Liberal Arts in Abbo of Fleury's Commentary on the Calculus of Victorius of Aquitaine." *Viator* 16 (1985), pp. 109–127; etc.
21 For editions of Gerbert's mathematical texts, see *Gerberti postea Silvestri II papae Opera Mathematica (972–1003)*. Ed. N. Bubnov. Berlin: R. Friedländer & Sohn, 1899 (repr. Hildesheim: Georg Olms, 1963).

first of the mathematical sciences: arithmetic, and its relationship to knowledge as a whole. At the same time, attention is, naturally, given to the way arithmetic is intertwined with the other sciences of the *quadrivium*. The sources used predominantly include Gerbert's correspondence, which mirrors a live dialogue that can point towards the themes that contemporary scholars (mostly Gerbert's disciples, friends, and colleagues) considered important or useful. Finally, the book does not ignore Gerbert's other texts and the works of his contemporaries and successors, in which we can track Gerbert's commitment to (especially) the field of arithmetic.

*

Given the fact that I am not a native English speaker, I am obliged to David Černín, Gary Frost, Igor Jelínek, Daniela Rywiková, and Světla Hanke Jarošová for their help with language aspects. Without their help, this book would not have existed in its current form. For invaluable discussions and advice, I am grateful to Daniel Špelda, Costantino Sigismondi, Seweryn Blandzi, and Richard Psík. Since many conclusions and parts of the text presented in this book have already been published as parts of my various studies (for more details, see the Bibliography), I would also like to thank all the anonymous reviewers who have provided priceless feedback. However, any shortcomings of this book are solely my own responsibility.

INTRODUCTION: ARITHMETIC AS THE SCIENCE OF NUMBERS

In the dialogue *Gorgias*, Plato, through the lips of Socrates, differentiates two kinds of science that are concerned with numbers. On the one hand, there is theoretical arithmetic, i.e., the art that focuses on the properties of numbers regardless of their concrete values. Thus, it is a theoretical discipline that inquires into the nature and properties of numbers or numerical relations, such as evenness or oddness, etc. Today, we may call it the theory or philosophy of numbers. On the other hand, there is practical arithmetic or counting, i.e., a very similar art that, contrary to the former, considers the specific values of numbers and inquires into these values. Thus, they are numerical operations proper; they form the content of mathematical education in schools and are known as basic arithmetical operations.[1]

In the seventh book of *The Republic*, Plato's Socrates differentiates both sciences of numbers once again; however, the primary goal is to show that both can elevate mankind, since they lead us to true knowledge, teach us rational competencies, and guide the soul from the everyday world subject to change to permanence and true being. Theoretical knowledge of numbers together with practical calculations allow us to grasp the very nature of numbers, which can elevate human reason towards true knowledge.[2]

During the Latin Christian Middle Ages, both sciences were given partially different roles. Theoretical mathematics was, following the ancient model, included among the liberal arts. Since late Christian antiquity, it has been accorded the most significant role in comprehending the way the world is ordered, in finding the traces of God's Will in created reality, in mankind's endeavour to know God, etc.[3] Arithmetic was also considered to possess metaphysical significance, since without numbers, nothing could exist, and no philosopher could be a philosopher without cultivating the mathematical (and especially the arithmetical) arts.[4]

---

1 Plato, *Gorg.* 451a–c; cf. also *Tht.* 198a–c.
2 Plato, *Resp.* VII, 8, 524d–526c.
3 A typical example is Aurelius Augustinus who praises the science of numbers in many places – see, for example, Augustine, *Ord.* II, 18, 48; or idem, *Lib. arb.* II, 8, 20–II, 12, 34.
4 Cf., e.g., Augustine, *Lib. arb.* II, 16, 42; Boethius, *Arith.* I, 1, p. 11; Isidore, *Etym.* III, 4, 1–4.

Continuity with ancient (especially Pythagorean) arithmetical tradition was established by Latin texts which, mostly in an elementary fashion, presented the basics of this art. The crucial Neopythagorean treatise,[5] which retained its influence from the turn of antiquity and the Middle Ages to the age of the Renaissance, was Boethius's loose translation of *Introduction to Arithmetic*[6] by Nicomachus of Gerasa. In contrast, counting (*computus*) was far more closely tied to practical life (merchant accounting, administration, etc.), and while it received some attention from scholars, it very seldom received the same appreciation as the theory and philosophy of numbers.[7]

---

5   Cf., for instance, Caiazzo, I., "Medieval Commentaries on Boethius's *De arithmetica*: A Provisional Handlist." *Bulletin de philosophie médiévale* 62 (2020), pp. 3–13; or Crialesi, C. V., "The Status of Mathematics in Boethius: Remarks in the Light of his Commentaries on the *Isagoge*." In: Giovannetti, L. (ed.), *The Sustainability of Thought: An Itinerary through the History of Philosophy*. Napoli: Bibliopolis, 2020, pp. 106–109.

6   There are numerous texts analysing Boethius's understanding of translation (method, aims, form, etc.) – for instance, Vogel, C., *Boethius' Übersetzungsprojekt. Philosophische Grundlagen und didaktische Methoden eines spätantiken Wissenstransfers*. Wiesbaden: O. Harrasowitz, 2016, especially pp. 125–169; Kárpáti, A., "Translation or Compilation? Contributions to the Analysis of Sources of Boethius' *De institutione musica*." *Studia Musicologica Academiae Scientiarum Hungariae* 29 (1987), pp. 5–33; Bower, C. M., "Boethius and Nichomachus: An Essay Concerning the Sources of *De institutione musica*." *Vivarium* 16/1 (1978), pp. 1–45; Vogel, C., "Die „boethianische Frage" – Über die Eigenständigkeit von Boethius' logischem Lehrwerk." *Working Paper des SFB 980 Episteme in Bewegung* 17 (2019), here pp. 23–24; Barnes, J., "Boethius and the Study of Logic." In: Gibson, M. (ed.), *Boethius. His Life, Thought and Influence*. Oxford: Basil Blackwell, 1981, p. 79; Moyer, A. E., "The *Quadrivium* and the Decline of Boethian Influence." In: Kaylor, N. H. Jr. – Phillips, P. E. (eds.), *A Companion to Boethius in the Middle Ages*. Leiden – Boston: Brill, 2012, p. 481, etc. Boethius's specific approach to translation was also discussed by authors of the Early Middle Ages – see, for example, Evans, G. R., "Introductions to Boethius's "Arithmetica" of the Tenth to the Fourteenth Century." *History of Science* 16 (1978), p. 26; or *De arith. Boeth.*, ad prol., p. 129.

7   In church and religious circles, the most significant role of counting can be seen in so-called *computus paschalis* or *computus ecclesiasticus*, i.e., the scientific discipline that served to compute the Easter holiday – see, e.g., Borst, A., *Computus: Zeit und Zahl in der Geschichte Europas*. Berlin: Wagenbach, 1990, pp. 24–43. A basic overview of the arithmetical pursuits during the Early Middle Age (around the 9th to the 12th century) is provided in the translations of representative texts with brief commentary by Folkerts, M. – Hughes, B., "The Latin Mathematics of Medieval Europe." In: Katz, V. J. (ed.), *Sourcebook in the Mathematics of Medieval Europe and North Africa*. Princeton: Princeton University Press, 2016, pp. 12–62.

The tendency to reconcile both sciences of numbers can be traced back, relatively easily, to the last quarter of the 10th century, when Abbo of Fleury wrote his aforementioned commentary on *Calculus* by Victorius of Aquitaine. The commentary dealt with summary numerical tables for multiplication, and introduces, alongside the arithmetical operations, the role of arithmetic in the contemporary system of knowledge.[8] The importance of arithmetic for natural philosophy, metaphysics, and theology is also emphasised.[9] At the same time, his contemporary Gerbert of Aurillac also exhibited a focused long-term interest in theoretical and practical arithmetic. Gerbert's commitment to both arithmetical fields brought to light several remarkable works that notably influenced the cultivation of these arts for centuries to come.

Therefore, this book will focus first on theoretical arithmetic, in which Gerbert not only creatively combines a philosophical approach to the importance of numbers in the human pursuit of true knowledge (especially the so-called *Saltus Gerberti*), but also thematizes the use of arithmetical knowledge in other *quadrivium* disciplines, inspired by Boethius's mathematical treatises. Subsequently, the arithmetical innovations for which Gerbert became famous (especially the abacus and the use of Western Arabic numerals) will be mentioned.

However, in order to properly present a comprehensive picture of Gerbert's arithmetical and counting arts of numbers, which on the one hand (as in the work of Theon of Smyrna,[10] or the aforementioned Nicomachus) fulfil the role of a philosophical propaedeutic that provide a metaphysical interpretation of reality, and on the other hand lead to the development of human intellectual abilities, widening cognitive capacities, we must briefly summarise the Early medieval view on arithmetic as a liberal art.

*

The position of arithmetic in early medieval Latin schools was in many ways exceptional. This integral part of the seven liberal arts was regarded as the first of the *quadrivium* (mathematical sciences) and its role was comparable to the role of logic (dialectics) within the *trivium* arts (language sciences).[11] Between

---

8 Cf., e.g., Abbo, *In Calc.* I, 1–II, 2, pp. 64–66.
9 For example, ibid. II, 3–16, pp. 66–72.
10 Cf., for example, Petrucci, F. M., "Theon of Smyrna: Re-thinking Platonic Mathematics in Middle Platonism." In: Tarrant, H. et al. (eds.), *Brill's Companion to the Reception of Plato in Antiquity*. Leiden – Boston: Brill, 2018, pp. 143–155.
11 Cf., e.g., Franci, R., "L'insegnamento dell'aritmetica nel Medioevo." In: Franci, R. – Pagli, P. – Rigatelli, L. T. (eds.), *Itinera mathematica. Studi in onore di Gino Arrighi per il suo 90° compleanno*. Siena: Università di Siena, 1996, pp. 112–113.

the late 4th and early 7th century, numerous Latin texts covered the subject, nature, and aims of arithmetic to varying degrees, and with differing objectives. Meanwhile, statements proposed in these texts became the authoritative basis for understanding arithmetical science for a large part of the Middle Ages. The cultivation of the subject matter of arithmetic at (especially early) medieval schools is therefore inconceivable without the works of Martianus Felix Capella, Aurelius Augustine, A. M. T. S. Boethius, F. M. A. Cassiodore, or Isidore of Sevilla, all of whom built their works upon the ancient foundations of the subject matter of arithmetic, and considerably aided the continuity of the cultivation of such knowledge in the Latin-Christian intellectual environment.

In the early Middle Ages, mathematics was mostly defined as a theoretical or speculative science (*scientia doctrinalis*).[12] This designation is based on the traditional Aristotelian division of the sciences, in which mathematics is ranked among the theoretical or speculative sciences occupying the middle ground between the 'first philosophy' (the highest science, i.e., metaphysics) and physics.[13]

Of all the early medieval scholars, Boethius was the leading medieval authority on the subject matter of mathematics and arithmetic, mainly due to his *Introduction to Arithmetic*.[14] Boethius also covered the exact position of mathematics in relation to philosophical knowledge in other texts. In the first commentary on Porphyry's *Isagoge*, he endorsed the Aristotelian division of philosophy by splitting it into theoretical or speculative philosophy (*theoretica, speculatiua, contemplatiua*) and practical philosophy (*practica, actiua*). Within

---

12 Cf., for example, Klinkenberg, H. M., "Divisio philosophiae." In: Craemer-Ruegenberg, I. - Speer, A. (eds.), *Scientia und ars im Hoch- und Spätmittelalter* I. Berlin: De Gruyter, 1994, pp. 3–19; Weisheipl, J. A., "The Concept of Scientific Knowledge in Greek Philosophy." In: Gagne, A. - De Koninck, T. (eds.), *Melanges a la Memoire de Charles De Koninck*. Quebec: Les Presses de l'Université Laval, 1968, pp. 487–507; idem, "The Nature, Scope, and Classification of the Sciences." In: Lindberg, D. C. (ed.), *Science in the Middle Ages*. Chicago: University of Chicago Press, 1977, pp. 461–482; etc.
13 Cf., for example, Aristotle, *Met.* XI, 7, 1064a–b.
14 Cf., for instance, Masi, M., "Boethius' *De institutione arithmetica* in the Context of Medieval Mathematics." In: Obertello, L. (ed.), *Atti del Congresso internazionale di studi Boeziani*. Roma: Editrice Herder, 1981, pp. 263–272; Kibre, P., "The Boethian *De Institutione Arithmetica* and the Quadrivium in the Thirteenth Century University Milieu at Paris." In: Masi, M. (ed.), *Boethius and the Liberal Arts*. Berne - Frankfurt a. M.: Peter Lang, 1981, pp. 67–80; or Guillaumin, J.-Y., "Boethius's De Institutione Arithmetica and Its Influence on Posterity." In: Kaylor - Phillips (eds.), *A Companion to Boethius…*, pp. 135–161.

theoretical philosophy, he established three fundamental scientific domains concerned with *intellectibilia* (divine science), *intellegibilia* (mathematics), and *naturalia* (physics).[15] Similarly, in the later theological treatise *De trinitate* (the whole title of which is *Quomodo trinitas unus Deus ac non tres dii*), written approximately twenty years after *Arithmetics*,[16] Boethius divides theoretical philosophy into physics (*disciplina naturalis*), mathematics (*mathematica*), and theology (*disciplina theologica*). The subject of mathematics is defined as abstracted from matter and motion, although it is present in matter as the forms (*formae*) of bodies.[17]

Mathematics is thus situated at the centre of theoretical philosophy. In contrast to physics, it focuses on something stable, although it is not as metaphysically exalted as theology (or metaphysics, i.e., the divine science), which inquiries into objects utterly independent of matter. From at least the time of Aristotle, the subject of mathematics had been defined as quantity which is divorced from matter – unchanging and stable, but existing in the material world.[18] This description was also accepted by Nicomachus.[19]

Cassiodore, Boethius's contemporary and successor in the highest Roman office of *magister officiorum*, who knew Boethius's works, including the translation of Nicomachus's *Arithmetic*,[20] expressed this theory very similarly. According to Cassiodore, mathematics is a theoretical science (*scientia doctrinalis*), since this science, out of all the speculative sciences, holds a subject of a theoretical nature in the highest degree (*excellentia*) of concern.[21] Mathematics deals with an abstract quantity (*quantitas abstracta*), that is, a quantity completely irrespective of the bearer of this quantity (abstracted from the material occurrence of

---

15 Boethius, *1 In Isag.* I, 3, pp. 8–9; cf. Hicks, A., *Composing the World. Harmony in the Medieval Platonic Cosmos*. Oxford: Oxford University Press, 2017, pp. 70–73.
16 See, for example, De Rijk, L. M., "On the Chronology of Boethius' works on logic II." *Vivarium* 2 (1964), pp. 125–126; Barnes, "Boethius…," p. 87; or Marenbon, J., *Boethius*. Oxford: Oxford University Press, 2003, pp. 76–77.
17 Boethius, *Trin.* 2, p. 8.
18 Aristotle, *De an.* III, 7, 431b; idem, *Phys.* II, 2, 193b–194a; or idem, *Met.* VI, 1, 1026a; and many others. For more details, see Studtmann, P., "Aristotle's Category of Quantity: A Unified Interpretation." *Apeiron* 37/1 (2004), pp. 69–91.
19 Nicomachus, *Arith.* I, 2, 5, p. 5.
20 Cassiodore, *Var.* I, 45, 4, p. 40; or idem, *Inst.* II, 4, 7, p. 140; cf. Caldwell, J., "The *De Institutione Arithmetica* and the *De Institutione Musica*." In: Gibson (ed.), *Boethius…*, pp. 137–138.
21 Cassiodore, *Inst.* II, 3, 21, p. 130.

quantity), so its subject is exclusively conceived either intellectually or rationally, while it ignores further detailed specification of that quantity.²²

Cassiodore's words are echoed by Isidore of Sevilla, who adds a proposition that can be understood as the further characterisation of the quantity. He includes in his delimitation of mathematics absence of other specifications of quantity, i.e., various criteria for different categories and types (for example, odds and evens, etc.).²³ Thus, quantity can be divided into two basic kinds: on the one hand, it is a multitude (*multitudo*), i.e., something firmly demarcated (*discreta*), delimited and countable, for example, individual trees, books, etc.; and, on the other hand, it is a magnitude (*magnitudo*), i.e., something continuous (*continua*), with a certain extent, and thus measurable – for example, the length of an item, the circumference of a sphere, etc.²⁴ Aristotle had already distinguished between discontinuous or disjunctive quantity (e.g., a number) and a continuous quantity (e.g., a line, or a solid),²⁵ that is, respectively, a quantity in which the individual parts themselves are not related (e.g., a number) and a quantity in which the parts are directly linked together (e.g., a line or a solid). The former is referred to as a multitude, while the latter is referred to as a magnitude.²⁶

Boethius analysed the distinction between multitude and magnitude in detail in a very similar manner in a passage about the category of quantity in his commentary on Aristotle's *Categories*²⁷ and (like Nicomachus) used the multitude–magnitude distinction to distinguish between the four special mathematical sciences, i.e., the *quadrivium*.²⁸ The multitude can be thought of in itself (*per se*), that is, as a discrete or delimited multitude, i.e., a number in itself, or as a multitude related to another multitude (*ad aliud, ad aliquid*), that is, when numbers are ordered according to numerical ratios. The former gives rise to

---

22 Ibid. II, praef., 4, p. 92.
23 Isidore, *Etym.* III, praef.
24 Nicomachus, *Arith.* I, 2, 4–5, pp. 4–5; Boethius, *Arith.* I, 1, p. 10; cf. Corti, L., "Scepticism, number and appearances. The ἀριθμητικὴ τέχνη and Sextus' targets in M I-VI." *Philosophie antique* 15 (2015), pp. 129–130. For an early medieval gloss, see, for instance, *In Boeth. Arith.* 15, p. 251.
25 Aristotle, *Cat.* 6, 4b.
26 Idem, *Met.* V, 13, 1020a.
27 Boethius, *In Cat.* II, c. 201C–205B; cf. Hicks, A., "Martianus Capella and the Liberal Arts." In: Hexter, R. – Townsend, D. (eds.), *The Oxford Handbook of Medieval Latin Literature.* Oxford: Oxford University Press, 2012, pp. 312–313.
28 This term for the quartet of liberal arts, which focus on mathematics and describe the world we are situated in, was used by Boethius for the first time – see Boethius, *Arith.* I, 1, p. 9, 11.

the doctrine of numbers, i.e., arithmetic, which inquiries into numbers *per se*, whereas the latter results in the science of music and musical intervals, whose subject are numerical ratios. The second kind of quantity, i.e., magnitude, can also be differentiated further. In this case, Boethius lists the criteria of stability (*immobilis*) and mobility (*mobilis*). The mathematical science that enquires into the unchanging and stable is geometry, while astronomy focuses on magnitudes in motion.[29]

In this way, the basic structure of mathematics emerges as it was mediated through Boethius's Nicomachean (i.e. Neopythagorean) reading: the subject of arithmetic is *multitudo per se*, geometry focuses on *magnitudo stabilis*, music deals with *multitudo ad aliquid,* and astronomy is concerned with *magnitudo mobilis*.[30]

According to Boethius, arithmetic enjoys the most important position among the other mathematical disciplines (*principium et mater*), since multitude *per se* is nothing other than the number itself, which is necessary for all other (not exclusively) mathematical sciences.[31] Without arithmetic, Boethius writes, there could be no geometry, music, astronomy, or any other kind of human knowledge at all. Boethius (following Nicomachus) proves the primacy of arithmetic by the following argument: Numbers (*numerus*) are an expression of God's thoughts according to which God created all of creation, therefore numbers must be antecedent (*prior*) by virtue of their nature (*natura*). When that which comes after (*posterior*) vanishes, for example, the species 'human' (*homo*), that which is antecedent, for example, the genus 'animal' (*animal*) is not affected; on the other hand, when that which is antecedent vanishes (animal), then all that comes after

---

29 Boethius, *Arith.* I, 1, pp. 10–11; Nicomachus, *Arith.* I, 3, 1–2, pp. 5–6; cf. Evans, G. R., "Boethius' Geometry and the Four Ways." *Centaurus* 25/2 (1981), pp. 161–165; or *Excerpt.*, p. 279.
30 Cf., e.g., Pizzani, U., "Il *Quadrivium* Boeziano e i suoi problem." In: Obertello (ed.), *Atti del Congresso…*, pp. 211–226; Masi, M., "The Liberal Arts and Gerardus Ruffus' Commentary on the Boethian *De Arithemtica*." *Sixteenth Century Journal* 10/2 (1979), p. 29; or Pizzani, U., "Studi sulle fonti del "De Institutione Musica" di Boezio." *Sacris erudiri* 16 (1965), p. 158; etc. For the reception of the quadrivium division in Middle Ages with illustrative diagrams, see Walden, D. K., "Charting Boethius: Music and the Diagrammatic Tree in the Cambridge University Library's *De Institutione Arithmetica*, MS II.3.12." *Early Music History* 34 (2015), pp. 207–228.
31 For early medieval adaptation, see Evans, "Introductions to Boethius's 'Arithmetica'…", p. 36; or *De arith. Boeth.*, ad I, I, pp. 132–133.

and is dependent on it (e.g., human) vanishes as well.[32] Arithmetic as the science of numbers thus precedes all other sciences, since nothing could exist without numbers.

Geometry, Boethius continues, needs arithmetic because it would not be possible to think about the forms (*formae*) of objects (e.g., triangle, quadrangle, etc.) without the ability to describe them using numbers. In music theory, i.e., numerical ratios (*proportiones*), it would not be possible to have various musical intervals (e.g., an octave, a perfect fourth, a perfect fifth, etc.), if there were no numbers, and, thus, music needs arithmetic. Astronomy would lack the ability to describe the orbits of celestial bodies (*circuli, centra*, etc.) and their distances and positions without the knowledge of geometry (geometrical shapes) and music (perfect celestial harmony, *armonica*, music of spheres); therefore, even in astronomy, numbers are a prerequisite. Without arithmetic there would be no geometry, no music, and, also, no astronomy.[33]

In this manner, Boethius establishes a certain hierarchy of the mathematical sciences. Arithmetic is necessarily the first among sciences, since it needs only numbers and nothing else to achieve its goals. Although geometry inquires into something *per se* (shapes), it needs numbers for its practice and, therefore, comes immediately behind arithmetic. Music does not focus on something *per se* – at the centre of its attention are the relative properties of numbers; therefore, it is also dependent on arithmetic (i.e., on numbers themselves) and comes third. Finally, astronomy needs both arithmetic and geometry, and cannot operate without music either; therefore, astronomy occupies fourth place among the mathematical sciences.[34]

---

32 Boethius, *Arith.* I, 1, p. 12; Nicomachus, *Arith.* I, 4, 1–3, pp. 9–10; cf. Fournier, M., "Boethius and the Consolation of the Quadrivium." *Medievalia et Humanistica. Studies in Medieval and Renaissance Culture* 34 (2008), p. 3; or *De arith. Boeth.*, ad I, I, p. 133.

33 Boethius, *Arith.* I, 1, pp. 12–14; Nicomachus, *Arith.* I, 4, 4–5, 3, pp. 10–11; cf. Guillaumin, "Boethius's De Institutione Arithmetica…," pp. 139–141. For early medieval reception, see *De arith. Boeth.*, ad I, I, pp. 134–136.

34 During late antiquity and in the Middle Ages, other methods of ordering the four mathematical sciences existed; e.g., Claudius Ptolemy understood astronomy to be the most important science not only among the mathematical sciences; he described it as more fundamental to human knowledge than metaphysics or theology, and, in some respects, as the most valuable of all the components of theoretical philosophy – see Ptolemy, *Alm.* I, 1, pp. 5–7. Another example of a different ordering of the mathematical sciences of the quadrivium is from the widely studied, commented on, and quoted Martianus Capella, who, in his work *On the Marriage of Philology and Mercury,* places geometry in first place and only after comes arithmetic, followed by astronomy and

Introduction: Arithmetic as the Science of Numbers

With explicit references to Plato, Boethius presented arithmetic as the necessary condition for all other sciences, since without numbers no science could function, including philosophy.[35] Holy Scripture also provides an excellent opportunity to elucidate the key role of numbers, which substantiate the Divine basis of the created world, since the *Book of Wisdom* states that God created everything in compliance with measurement, number, and weight (*mensura, numerus, pondus*).[36] Numbers form the basis of the order built by God, according to which everything is arranged in the universe. Thus, numerical ratios are those maintaining harmony within the created world and, at the same time, represent the harmonious relationship between the Creator and the creation.[37]

Therefore, since arithmetic is always primarily concerned with numbers, it can rightly be called the science of numbers. Indeed, Cassiodore in *Institutiones* explains that arithmetic is an appropriate name, since it deals with numbers,[38] and Isidore of Sevilla repeats this when he writes that arithmetic is the doctrine of numbers because the Greeks call the number ἀριθμός.[39] All this implies that the subject of arithmetic is the number as quantitative determination.

From antiquity, the idea has existed that every quantity can be expressed by a unit or number.[40] Each number is derived from the unit, so the unit can be described as the cause of the number, the mother of all numbers, their essential source.[41] Thus, the unit cannot be regarded as a number, and the relationship between numbers and unit is similar to the relationship between the Creator and the Creation. Boethius says that 'One' is the foundation of God and the absolute Good, from which dualities immediately arise (the number 'two') – for example,

---

music – see Martianus, *De nupt.* VI, 567–IX, 996, pp. 201–384. Other similar examples could be listed – cf. Teeuwen, M., "The Pursuit of Secular Learning: The Oldest Commentary Tradition on Martianus Capella." *Journal of Medieval Latin* 18 (2008), pp. 36–51; or Stahl, W. H., *The Quadrivium of Martianus Capella: Latin Tradition in Mathematical Sciences 50 B.C.–A.D. 1250. Martianus Capella and the Seven Liberal Arts.* Vol. 1. New York – London: Columbia University Press, 1971.

35 Boethius, *Arith.* I, 1, p. 11. Cf. Plato, *Resp.* VII, 6, 522c.
36 *Sap.* 11,20. Cf., for example, Augustine, *Gen. ad litt.* II, 1.
37 Boethius, *Arith.* II, 1, pp. 93–94.
38 Cassiodore, *Inst.* II, 4, 2, p. 133.
39 Isidore, *Etym.* III, 1, 1. The translation of the Greek term for number into Latin by the term *numerus* was common in the early Middle Ages, and Aurelius Augustinus frequently operated with it in his works – see, e.g., Augustine, *Ord.* II, 14, 40; or idem, *Mus.* III, 1, 2.
40 For example, Aristotle, *Met.* X, 1, 1052b.
41 Cf., e.g., Boethius, *Arith.* I, 17, p. 45.

darkness and light, Heaven and Earth, etc. If everything is related to its root cause (i.e., if everything is contained in a number, or determined by a numerical ratio), then it also participates in the supreme Good. The moment anything deviates from numbers, it also turns away from good, units, and thus from God himself.[42]

Nicomachus of Gerasa in *Introduction to Arithmetic* mentions the three most commonly used definitions of the number – numbers can be described as a discrete (limited) quantity, or as a combination of units, or as an infinite set that originates from the unit (and apparently returns to it).[43] Although these definitions may seem, at first glance, quite different, we can find a unifying thread running through them which also corresponds with the late ancient and early medieval interpretations of arithmetical subject matter – that is, with contents of treatises on the mathematical properties of numbers.

The most widely used definition of the number in the early Middle Ages was Nicomachus' second assessment, which states that the number is a combination or arrangement of units. For example, Boethius, in his loose translation of the Nicomachean treatise, indicates that numbers are a collection of units.[44] Similarly, other texts from the epoch speak of the number as the arrangement or orderings (*congregatio, compositio, constitutio*) of units (*monades, unitates*).[45] By this definition a specific number can be imagined as a group of units. This definition was already prevalent in antiquity. According to Iamblichus, we can find the origins of such a designation of the number in the works of Thales of Miletus,[46] although Aristotle's claim that numbers were defined in such a way by the Pythagoreans might be regarded as more probable.[47] The ancient vogue for perceiving numbers in this manner is evidenced by the presence of the same definition in Euclid's *Elements*.[48]

Defining number as a collection of units gives rise to the idea of the figural representation of numbers. The ordering of units, or representation of them using points (*puncti*) in geometrical shapes, corresponds to natural (*naturalis*) signs

---

42 Ibid. I, 32, p. 80.
43 Nicomachus, *Arith.* I, 7, 1, p. 13.
44 Boethius, *Arith.* I, 3, p. 15; cf., e.g., *Excerpt.*, p. 279.
45 Cf. Martianus, *De nupt.* VII, 743, p. 269 (*congregatio monadum*); Cassiodore, *Inst.* II, 4, 2, p. 133 (*ex monadibus multitudo composita*) or Isidore, *Etym.* III, 3, 1 (*multitudo ex unitatibus constituta*).
46 Iamblichus, *In Nic. Arith.* p. 10.
47 Aristotle, *Met.* I, 5, 986a.
48 Euclid, *Elem.* VII, def. 2, p. 103.

for numerical values, in contrast to Greek (Nicomachus) or Roman (Boethius) numerals, and (we could add) also to Arabic numerals. While these commonly used numerical symbols (*signa numerorum*) are instituted by humans, figurate numbers show numbers as sets of units, i.e., if the number '5' is expressed, then this numeral symbol does not correspond to the natural character of the value 'five', as this can be achieved only by an ordering of five units together.[49]

The insight that numbers are collections of units ordered into certain geometrical shapes reveals the direct relation of numbers to the created world surrounding human beings, whose building blocks are geometrical forms. When units are ordered into lines, we get linear numbers (*numeri lineares*) characterized by longitude (*longitudo*) as the only direction or dimension (*interuallum*). When the points are ordered in two directions (besides a longitude, there is also a latitude, *latitudo*), plane numbers arise (*numeri plani*), e.g., triangular numbers (*triangulares*), tetragonal numbers (*quadrati*), pentagonal numbers (*pentagoni*), etc. When a third dimension, altitude (*altitude*), is added to length and width, we get solid numbers (*numeri solidi*), that is, pyramidal numbers (*pyramides*), cubic numbers (*cybi*), etc. These types of numbers are similarly characterized by Boethius and Nicomachus, especially in the way they establish that figurate numbers refer to various numerical ratios.[50] The relationship to geometry is obvious here.

It is probably not a coincidence that the first definition by Nicomachus represented the number as a quantity specified in a certain way. Similarly, early medieval texts on arithmetic do not start by explaining the subject matter of arithmetic through figurate numbers, but by using several classifications and typologies of numbers. This subject matter, by its nature, seems to correspond precisely with the definition of numbers which Cassiodore gives in *Institutiones*,

---

49 Boethius, *Arith.* II, 4, pp. 106–107; Nicomachus, *Arith.* II, 6, 2–3, pp. 83–85.
50 Boethius, *Arith.* II, 5–39, pp. 110–172; Nicomachus, *Arith.* II, 6, 4–20, 5, pp. 85–119; for a brief interpretation, see Guillaumin, "Boethius's De Institutione Arithmetica…," pp. 149–151; for reception in medieval education, see, for example, Høyrup, J., "Mathematics Education in the European Middle Ages." In: Karp, A. – Schubring, G. (eds.), *Handbook on the History of Mathematics Educations*. New York: Springer, 2014, especially pp. 111–112; cf. also *De arith. Boeth.*, ad II, IIII–II, XXXII, pp. 144–147; or *In Boeth. Arith.* 64–81, pp. 260–263. Euclid also deals with this issue, but only marginally, and, to a certain extent, in a different manner – cf., e.g., definition of plane (and then square) numbers or solid (and then cube) numbers – Euclid, *Elem.* VII, def. 16–20, p. 104.

in which he states that numbers may be understood as a discrete quantity (*quantitas discreta*).⁵¹

In this respect, numbers are a specific quantity. Each number is explicitly shown in a particular delimitation, for example, the number 'five' is a quantity which is limited to the value of five. Each number is therefore fixed and has nothing in common with another number. The source of Cassiodore's definition of the number was probably Boethius's *Introduction to Arithmetic*, in which he states that numbers are always discrete (*discreta*) and may enter into mutual relationships as such (ratios), creating the orderliness and harmony of the world.⁵² Several ancient authors also touched on this possible definition of a number, for example, Aristotle in *Categories* or *Metaphysics*,⁵³ or (according to later references from Iamblichus) Eudoxus of Cnidus.⁵⁴

The aforementioned explanation clearly reflects the traditional understanding of the definition.⁵⁵ If mathematics is a genus superior to individual mathematical arts (including arithmetic), and if the subject of mathematics is abstract quantity, then the subordinate kind of mathematics, meaning here arithmetic, also consistently takes quantity as its subject – discrete as opposed to indiscrete (abstract). In this case, the quantity plays the role of a superior genus, and the definition of such a quantity represents a specific difference.

Quantities may be determined in various ways. It has already been suggested that the given definition may determine the actual value of a specific number. However, quantities can also be determined by other criteria that enable the creation of all manner of classifications of numbers. In particular, the typology of numbers represents the subject matter of arithmetic, preliminarily solved in late antiquity.

The most frequent methods of classifying numbers, which presumably appeared in all the early medieval texts on arithmetic, are the classifications of numbers into odd (*impar*) and even (*par*), and their subtypes (even times even, even times odd, odd times even, and optionally also odd times odd; respectively, into primes, and composite or intermediate numbers), or into superfluous (*superflui*), diminutive (*deminuti*), and perfect numbers (*perfecti*).⁵⁶

---

51 Cassiodore, *Inst.* II, 4, 2, p. 133.
52 Boethius, *Arith.* I, 2, p. 15.
53 Aristotle, *Cat.* 6, 4b; or idem, *Met.* V, 13, 1020a.
54 Iamblichus, *In Nic. Arith.*, p. 10.
55 Cf., for example, Aristotle, *Top.* I, 5, 101b.
56 Cf., e.g., Nicomachus, *Arith.* I, 7–16, p. 13–44; Boethius, *Arith.* I, 3–20, p. 15–54; cf. Guillaumin, "Boethius's De Institutione Arithmetica…," pp. 142–145. These issues are mentioned also in the most influential and best-known work of ancient mathematics,

Introduction: Arithmetic as the Science of Numbers

The numerical properties determined by the given identifications of quantity, that is, the typological traditions of late antique arithmetic, have been used widely since the early Middle Ages to explain the significance of certain numerical values that play an essential role, for example, in the Bible.[57] Above all, we should mention Saint Augustine, who, for instance, in *De civitate Dei*, among other things, explains why the act of creation took six days.[58] The number 'six' is a perfect number, as it is equal to the sum of its divisors, that is, 'six' can be completely divided by the numbers 'three', 'two' and 'one', and the sum of these divisors is equal to the number 'six'. Augustine adds that similar numbers are very rare, as most numbers are diminutive (the sum of the denominators, respectively divisors, is smaller than the value of the divided number – e.g., the number 'nine' or 'ten') or superfluous (the sum of the denominators, respectively divisors, is greater than the value of the divided numbers – e.g. the number 'twelve').[59] As Nicomachus states, with diminutive numbers we experience the failure of their parts to create the original unit, while in the case of superfluous numbers, the situation in which their parts create more than the sum of the original unit occurs.[60] Only in the case of perfect numbers do their parts add up to a whole (in the early Middle Ages, only the first four perfect numbers were used: 6, 28, 496 and 8128, although the algorithm for finding others was known).[61] For Augustine,

---

    i.e., Euclid's *Elements*. Although Euclid defines (some) properties of numbers according to the listed classificatory theories as well, in many cases he adopts a different stance and follows different goals – cf. Euclid, *Elem.* VII, def. 6–14, 22, pp. 103–105; therefore, for the medieval (at least until the 12th century) reception of arithmetic as a science, his work (although partially translated into Latin by Boethius) was not relevant, and its influence on medieval arithmetic can be considered marginal in comparison to Nicomachus, and in contrast to geometry, in which Euclid's thinking can be traced more clearly – cf. Stahl, W. H., *Roman Science. Origins, Development, and Influence to the Later Middle Ages*. Madison: University of Wisconsin Press, 1962, pp. 198–201; or Masi, M., "Arithmetic." In: Wagner, D. L. (ed.), *The Seven Liberal Arts in the Middle Ages*. Bloomington: Indiana University Press, 1983, pp. 162–164. For early medieval reception in commentary and glosses see, e.g., *De arith. Boeth.*, ad I, IIII–I, XX, pp. 136–139; or *In Boeth. Arith.* 15–40, pp. 251–255; etc.

57 For more details see, e.g., Meyer, H. – Suntrup, R. (edd.), *Lexikon der mittelalterlichen Zahlenbedeutungen*. München, W. Fink, 1987.
58 Augustine, *Civ. Dei* XI, 30.
59 Ibid.
60 Nicomachus, *Arith.* I, 14–15, p. 36–39.
61 Cf., for example, Boethius, *Arith.* I, 20, p. 51.

the arithmetical perfection of the number 'six' is what determined the number of days in which God realized the fundamental act of creation.

In the next chapter of *De civitate Dei*, Augustine moves on to the number 'seven', marked as the number of completion since it corresponds to the six days of creation and the seventh day of rest. Therefore, 'seven' can be appropriately considered a reference to everything finished and created, that is, everything that has its origin and its harmony in numbers or numerical ratios. Therefore, this figure can be conceived as the symbol of all numbers. The reasons for the designation of 'seven' as the expression of all numbers is not only down to biblical allusions, it is also justified by the arithmetical properties of the numbers. The number 'seven' is, in fact, the sum of two numbers, which together represent all numbers – odd numbers are represented by the lowest odd number (i.e., 'three'), and even numbers by the lowest even (and, at the same time, even times even) number, i.e., 'four'.[62]

The pre-eminence of the number 'three' among odd numbers is probably universally understood (the number 'one' is not a number, the number 'two' is not odd), but in the case of the number 'four', doubts may arise, as the lowest even number might be regarded as the number 'two'. However, this understanding of the number 'two' was problematic in several ways to late antiquity and the early Middle Ages.[63] For example, odd numbers are usually defined as being impossible to divide into two equal integers, since one component will always be a unit larger or smaller. Even numbers are thus defined as those that can be divided into two equal integers.[64] Neither of these definitions is valid for the number 'two' – unlike odd numbers, it may be divided into two identical components, but not into the same two numbers, as required by the definition of even numbers (since a unit is not a number, but a source of numbers). Should anyone still maintain that the number 'two' is an even number because it can be divided into two equal halves, another problem soon arises. In the typology of even numbers, the number 'two' would match the definition of an even times even number (i.e., such numbers that can always be divided into equal halves until we reach a unit, namely the numbers 4, 8, 16, 32, etc.) and the definition of an even times odd number (although such even numbers can be divided into two equal halves, this

---

62 Augustine, *Civ. Dei* XI, 31.
63 For more details see, for instance, Radke, G., *Die Theorie der Zahl im Platonismus: Ein systematisches Lehrbuch*. Tübingen: A. Francke, 2003, pp. 438–446.
64 Cf., e.g., Boethius, *Arith.* I, 3, p. 16; Nicomachus, *Arith.* I, 7, p. 13; or Isidore, *Etym.* III, 5, 2 etc. See also different definition by Euclid, *Elem.* VII, def. 8, p. 103.

division does not result in an even number, so it cannot be further divided into equal halves – these are, for example, the numbers 6, 10, 14, 18, etc.).[65] For these reasons, the number 'two' is often considered a somewhat strange and controversial figure that cannot be safely placed in the usual categories of discrete quantity; thus, the first really definite even number becomes the number 'four'.

The third definition of numbers was most comprehensively developed by Martianus Capella, who, in addition to the widespread definition of numbers as a collection of units, also states that numbers can be understood as a certain multitude with its source in the unit to which it returns.[66] He also included this method of characterizing numbers in his description of Lady Arithmetic, who appears in front of the gods during a wedding congregation, and whose appearance frightens the celestials. One of the main causes of their terror is the mysterious, barely visible ray (*radius*) that emanates from the venerable Arithmetic's forehead (*frons*). It then spreads out and expands before shrinking again and eventually returning to its source.[67] This mysterious ray is the image of numbers with their origins in units. All the multitude of numbers is dependent on this source, whereas individual numbers are connected to each other (mainstream ray) but can also enter into various other relationships (branching rays) in which there are fixed relationships between the numbers depending on the previous figures and the relationships between them. On the one hand, numbers originate from a single source and may proceed to infinity, but they can return to their primary source and the mother of all numbers (the shrinking of the beam).

This definition of numbers also appears in Boethius[68] in a partially modified form, whose ancient origins can be still be traced, while John Stobaeus recorded extracts from the works of the Neo-Pythagorean thinker Moderatus of Gades, which suggest that numbers are what emerge from the unit and return to it again.[69]

This definition of numbers draws attention to the infinite series of numbers and to the interdependence and dependence of numbers on each other. The given definition indicates that numbers not only have properties of their own but also receive certain characteristics thanks to their relations to other numbers. It can be assumed that this definition refers, among other things, to the next big

---

65 Cf., e.g., Nicomachus, *Arith.* I, 8–9, pp. 14–21; Boethius, *Arith.* I, 9–10, pp. 21–30; or Isidore, *Etym.* III, 5, 3–4. See also differently Euclid, *Elem.* VII, def. 9, p. 103.
66 Martianus, *De nupt.* VII, 743, p. 269.
67 Ibid. VII, 728–729, pp. 260–262.
68 Boethius, *Arith.* I, 3, pp. 15–16.
69 Stobaeus, *Ecl.* I, 1, 8, p. 5.

topic of contemporary theoretical arithmetic: the properties of numbers, insofar as they are related to other numbers – that is, the relative properties of numbers (mainly numerical ratios). Ancient and medieval arithmetic, in this regard, distinguishes those numbers that have the same value (*aequalis*, e.g., a dozen and a dozen, an ell and an ell, etc.) from those that do not have the same value (*inaequalis*, e.g., a dozen and threescore, a foot, and an ell, etc.).[70] Unequal numbers are then divided into those in which a larger number is compared to a lesser (ratios derived from multiples: that is, multiples, superparticular numbers, superpartient numbers, superparticular multiples and superpartient multiples) and numbers in which a smaller number is compared to a larger (ratios derived from divisors: that is, divisors, subsuperparticular numbers, subsuperpartient numbers, subsuperparticular divisors and subsuperpartient divisors).[71]

All these ratios arise from the equality provided by unity (the 1 : 1 ratio). It creates the order and rules that are present in this world, it enables (in compliance with fixed mathematical rules) the creation of all ratios, and, when reversed, it also points the way back to unity and equality – the goal of all created things. Thus, numerical sequences determined by a specific ratio are at the forefront of arithmetical interest, which is traditionally the apex of arithmetical learning – arithmetical, geometrical and harmonic proportion.[72]

---

70 For more details, including the reception in the Middle Ages, see, e.g., Albertson, D., "*Boethius Noster:* Thierry of Chartres's *Arithmetica* Commentary as a Missing Source of Nicholas of Cusa's *De docta ignorantia*." *Recherches de Théologie et Philosophie médiévales* 83/1 (2016), pp. 143–199.

71 Cf., e.g., Nicomachus, *Arith.* I, 17–II, 5, pp. 44–82; Boethius, *Arith.* I, 21–II, 3, pp. 54–105, or Isidore, *Etym.*, III, 6, 1–13. See also Euclid, *Elem.* VII, def. 3–4, p. 103. For a brief interpretation, see Guillaumin, "Boethius's De Institutione Arithmetica…," pp. 145–148; cf. *In Boeth. Arith.* 41–63, pp. 256–260; or *De arith. Boeth.*, ad I,XX–II,II, pp. 139–144; for an analysis of its influence on medieval music theory, see Crossley, J. N., "The Writings of Boethius and the Cogitations of Jacobus de Ispania on Musical Proportions." *Early Music History* 36 (2017), pp. 14–24; Rimple, M. T., "The Enduring Legacy of Boethian Harmony." In: Kaylor – Phillips (eds.), *A Companion to Boethius…*, pp. 448–449, 453; or Masi, M., "The Influence of Boethius *De Arithmetica* on Late Medieval Mathematics." In: Masi (ed.), *Boethius and the Liberal Arts*, pp. 81–95.

72 Cf. Nicomachus, *Arith.* II, 21–27, pp. 119–140; or Boethius, *Arith.* II, 40–50, pp. 172–213. See also different definition of proportion in Euclid, *Elem.* VII, def. 21, pp. 104–105. For a brief interpretation, see Guillaumin, "Boethius's De Institutione Arithmetica…," pp. 151–153; and for reading and influence in the Middle Ages, see, for instance, *In Boeth. Arith.* 82–112, pp. 264–269; *De arith. Boeth.*, ad II,XL–II,LII, pp. 147–150; or Crossley, "The Writings of Boethius…," pp. 13–15.

It seems that early medieval understanding of the subject matter of arithmetic is unambiguously linked to the Neo-Pythagorean (Nicomachean) tradition of the cultivation of this science. While the subject of mathematics is an undetermined abstract quantity, the various mathematical sciences define quantity in a certain way. To create this system, Boethius deployed distinctions such as *multitudo* and *magnitudo*, *per se* and *ad aliud*, respectively, *stabilis* and *mobilis*; he thereby defined and hierarchically organized the four basic mathematical sciences: arithmetic, geometry, music, and astronomy.

The very subject of the first mathematical science, i.e., arithmetic, is thus marked by the concretization (definition) of quantity into numbers. However, in the early Middle Ages, more definitions of numbers were used, not necessarily determined only by different approaches to the specification of the nature and essence of numbers. The reasons behind the different definitions are a topic in their own right, and were discussed in the context of the early medieval compendia on arithmetic and texts inquiring into arithmetic.

If numbers are characterized as a discrete quantity (*quantitas discreta*), then it highlights the direct link to the basic mathematical properties of numbers and their typology (i.e., the first major topic that early arithmetic dealt with). If numbers are defined as the sum of their units (*collectio unitatum*), it suggests the idea of figurate numbers – the second broad topic of early medieval textbooks on arithmetical knowledge. Finally, numbers might be understood as a stream that springs from the beginning (*a monade veniens*), which can gradually expand to infinity, which can branch out, but which can also eventually return to its source (*in monade desinens*). This conception fully corresponds to the issue of how numbers are related to other numbers – that is, to the relative properties of numbers and numerical ratios that establish numerical sequences etc., which is the last, major topic of theoretical arithmetic as cultivated in (early) medieval schools.

I. THEORETICAL ARITHMETIC

As mentioned above, from the time of the Pythagoreans, it was a well-known fact in philosophical debates that mathematics and numbers were not only the tools to capture the quantitative aspects of reality but were also essential parts of ontology, metaphysics, theology, epistemology, anthropology, cosmology, and other related fields of research. The Divine Unit, the precise and perfect order determining natural events, or the music of the spheres represent just a brief selection of the topics that made an art out of the science of numbers, an art linked to the very essence of the universe.

This part of the book focuses on theoretical arithmetic and the philosophy of numbers, which served as an essential tool for the metaphysical understanding of reality in many different ways. Attention will primarily focus on Gerbert's correspondence with his friends, disciples, and colleagues, in which they discuss contemporary issues of the *quadrivium*. First, Gerbert's letter to Constantine of Fleury and Micy is analysed. This letter, sometimes called *Saltus Gerberti*, deals with the somewhat delicate issue of converting three-member numerical sequences, which is clearly source-related to Boethius's *Introduction to Arithmetic*. An exposition of Gerbert's letter and, in particular, his method, which he proposed as an interpretation of the methodical instructions from the opening chapter of the second book of Boethius's arithmetical work, will be framed by contemporary debates regarding this issue at the end of the 10th century, i.e., especially in contrast with texts by Notker of Liège and Abbo of Fleury. Since the number is characterised (according to the third definition mentioned in the introduction) as a stream of quantity that springs from the unit (and goes back to it), representing the act of creation by God, the whole issue of numerical sequences following a given ratio, their origin, and conversions are seen as a description of the tools God used to create the world, and which corresponds to the hierarchy of the created world.

In the following chapters, the importance of arithmetic to other sciences of the *quadrivium* will be examined. The differences in the expression of numerical values between arithmetic and geometry will be illustrated by the example of Gerbert's late letter to Adelbold of Utrecht, in which the author explains, through triangular numbers, the differences between figural numbers (arithmetic) and the area of a triangle (geometry). The other two letters by Gerbert to Constantine of Fleury will comprehensively demonstrate the necessity of arithmetical knowledge for a proper understanding of music (both texts are essentially commentaries to Boethius's *Introduction to Music*).

At the of end of this part of the book, two letters from Gerbert will be analysed – the first was addressed to his otherwise unknown brother Adam and the second to Constantine of Fleury and Micy once again. These letters deal with the applied knowledge of numbers with regard to astronomy and they are related to Gerbert's theories concerning timekeeping. In one case, the subject of inquiry is a construction manual for an observational hemisphere (the so-called *De sphaera* treatise), while the second letter is horological in focus, demonstrating the changes in the presence of the Sun above the horizon throughout the year. We conclude by examining Gerbert's possible activities related to the construction of timekeeping devices.

1. The Nature of Numbers: *Saltus Gerberti* (Letter to Constantine)

Let us now deal with Gerbert's letter to Constantine, in which the author attempts to – in accordance with the nature of numbers (*natura numerorum*) – expound how to follow Boethius's *Introduction to Arithmetic* and how to convert three-member numerical sequences so that the multitude of numbers and numerical relations may adequately lead to their origin, i.e., the unit or the ratio 1 : 1, that is – equality. However, first, it is necessary to briefly contextualise Gerbert's letter in his life and work. The origins of Gerbert's scholarly reputation can be dated to the 970s, when he began to cooperate with the Ottonian imperial dynasty and became a teacher in Reims. Gerbert's contemporaries marvelled especially at his knowledge of the *quadrivium* and his unusual emphasis on the practical use of individual pieces of knowledge: the simplification of geometric and arithmetical calculations by means of a counting table (abacus), the use of observational astronomical instruments, the construction of timekeeping mechanisms, and his innovations to the contemporary organ and related instrumental techniques, etc.[1] He seems to have mastered the liberal arts – especially

---

1 See, e.g., Richer, *Hist.* III, 55, pp. 198–199; cf. Darlington, "Gerbert, the Teacher", pp. 456–476; Lindgren, U., *Gerbert von Aurillac und das Quadrivium. Untersuchungen zur Bildung im Zeitalter der Ottonen*. Wiesbaden: Steiner Verlag, 1976; DeMayo, C., "The Students of Gerbert of Aurillac's Cathedral School at Reims: An Intellectual Genealogy." *Medieval Prosopography* 27 (2012), pp. 97–117; Schärling, A., *Un portrait de Gerbert d'Aurillac: Inventeur de l'abaque, utilisateur précoce des chiffres arabes et pape de l'an mil*. Lausanne: Presses polytechniques et universitaires romandes, 2012; Williams, "The Cathedral School…," pp. 661, 675–676; etc.

during his approximately three-year stay (967–970) on the Iberian Peninsula – unusual for the given period in the Latin West.[2]

During Gerbert's teaching in Reims, there was a monk called Constantine of Fleury and Micy among his pupils or collaborators.[3] In the second half of the 970s (or early 80s), Gerbert dedicated several letters and essays to him, including the interpretation of certain passages of Boethius's *Introduction to Arithmetic II, 1*, the so-called *Saltus Gerberti*, i.e., *Gerbert's Leap*, or (according to Bubnov's title) *Scholium ad Boethii Arithmeticam Institutionem I. II, c. 1*. In this text, most likely written between 978 and 980,[4] Gerbert attempts to clarify a process, not fully explained by Boethius, in which unequal ratios between numbers are converted to equality and identity.

Another mathematical treatise was written slightly later by Abbo of Fleury, who, after his studies (e.g., in Paris or Reims) in the 980s, worked as a teacher in a convent school in Ramsey.[5] Probably in the first half of

---

[2] See Ademar, *Chron.* III, 31, p. 154; Richer, *Hist.* III, 43, pp. 191–192; William, *Gesta reg.* II, 167, 1–3, p. 280; cf. for instance, Sigismondi, C., "Gerberto, gli Arabi e Gerusalemme." *GERBERTVS – International Academic Publication on History of Medieval Science* 1 (2010), pp. 270–294; Zuccato, M., "Gerbert of Aurillac and a Tenth-Century Jewish Channel for the Transmission of Arabic Science to the West." *Speculum* 80 (2005), pp. 742–763; Samsó, J., "Cultura científica àrab i cultura científica latina a la Catalunya altmedieval: El monestir de Ripoll i el naixement de la ciència catalana." In: Udina i Martorell, F. (ed.), *Symposium internacional sobre els oríges de Catalunya (segles VIII–XI)*. Vol. 1. Barcelona: RABL, 1991, pp. 253–269; Udina i Martorell, F., "Gerberto y la cultura hispanica: los Manuscriots de Ripoll." In: Tosi (ed.), *Gerberto – scienza, storia e mito…*, pp. 35–50.

[3] For more details see, e.g., Warren, F. M., "Constantine of Fleury, 985–1014." *Transactions of the Connecticut Academy of Arts and Science* 15 (1909), pp. 285–292; or Head, T., "Letaldus of Micy and the Hagiographic Traditions of the Abbey of Nouaillé. The Context of the *Delatio corporis S. Juniani.*" *Analecta Bollandiana* 115/3-4 (1997), pp. 253–267.

[4] For datation see *The Letters of Gerbert with His Papal Privileges as Sylvester II*. Transl. H. P. Lattin. New York: Columbia University Press, 1961, p. 39; cf. also Gerbert d'Aurillac, *Correspondance*. Ed. & transl. P. Riché – J.-P. Callu. Paris: Les Belles Lettres, 2008, p. 693; or Gerberto, *Epistolario*. Transl. M. G. Panvini-Carciotto – C. Sigismondi – P. Rossi. Roma: APRA, 2010, p. 193.

[5] Cf., for example, Burnett, C., "King Ptolemy and Alchandreus the Philosopher: The Earliest Texts on the Astrolabe and the Arabic Astrology at Fleury, Micy and Chartres." *Annals of Science* 55 (1998), p. 332; or Lutz, *Schoolmasters…*, pp. 43–44.

the 980s,[6] he commented on *Calculus* by Victorius of Aquitaine, which he called by the "biblical name" *De numero, mensura et pondere*.[7] This commentary to Victorius's preface to subsequent tables is, therefore, directly linked to Boethius's *Introduction to Arithmetic*, whose purpose and goal is directed elsewhere. Nevertheless, Abbo refers to Boethius's authoritative text several times.[8] Above all, he discusses the issues that the aforementioned introductory chapter of the second book of *Introduction to Arithmetic* is concerned with.

Although Gerbert writes his *Scholium* as an interpretation of Boethius, while Abbo's text is instead a commentary on Victorius, they both allude to the fact that the transfer of ratios expressing the inequality of numbers needs to be appropriately executed, avoiding confusion (*non confuse, sed ordinate*).[9] It seems that they both respond to a specific unsatisfactory method of creating transitions between numerical sequences that was practised at the time. However, since neither of the authors is explicit on this point, we can only speculate.

From approximately the same period, another brief description of how to convert each numerical sequence to its original equality has survived. This text is called *De superparticularibus*, and its authorship is attributed to Notker of Liège.[10] There is no doubt that Abbo and Gerbert were familiar with the interpretation of transitions between sequences usually associated with Notker, which shows clear features of the confusion they both mention. Thus, there might be genuine connections between all three texts, as suggested in the extant notes from the 12th century.[11]

---

6 For datation see Peden, A. M., "Introduction." In: *Abbo of Fleury and Ramsey, Commentary on the Calculus of Victorius of Aquitaine*. Ed. A. M. Peden. Oxford: Oxford University Press – The British Academy, 2003, p. xiv.

7 Abbo, *Quaest. gram.* 50, p. 275; or idem, *In Calc.* II, 1, p. 65.

8 E.g., Abbo, *In Calc.* III, 24, p. 87.

9 Gerbert, *In Boeth. Arith.* 1, p. 34; cf. Abbo, *In Calc.* III, 22, p. 86.

10 Notker, *Superpart.*, pp. 297–299; see also Caiazzo, I., "Un commento altomedievale al De arithmetica di Boezio." *Archivum Latinitatis Medii Aevi* 58 (2000), p. 118, 125; or Evans, G. R., "The *Saltus Gerberti*: The Problem of the 'Leap'." *Janus* 67 (1980), pp. 264–266.

11 Evans, "The *Saltus Gerberti*…," p. 267.

Notker's works have been preserved in a manuscript from the monastery in Tegernsee (now Munich, Bayerische Staatsbibliothek, CLM 18764, fols. 78v–79r), serving as a postscript to Boethius's *Introduction to Arithmetic*. Part of this manuscript is also a preface (*accessus*) and commentary to Boethius's manuscript called *De aritmetica Boetii*, edited by I. Caiazzo.[12] These texts most likely originated in the late 10th century,[13] thus creating an authentic complement to the three previously mentioned treatises.

A subject of interest for authors in the late 10th century was the main introductory passage of the second book of Boethius's *Arithmetics*, in which the author claims that any inequality derives from former equality, analogous to the concept that everything material is composed of four elements (*elementa*), words are composed of letters (*litterae*), and music is based on sounds (*sonus*). All these components stem from a former identity, equality, and unity.[14] Since everything is created according to numerical ratios,[15] and God Himself is originally one and the same undivided being,[16] it is clear that any proportion or ratio has its roots in unity and equality.[17]

Equality and inequality in Boethius's *Arithmetic* are defined as two types of relative properties of numbers, i.e., such properties of numbers in which the relationship between their numerical values is considered.[18] In the case of equality, the compared values are not smaller or larger, but identical, i.e., they are the same size, as if we were comparing ten and ten, three and three, a cubit and a cubit, a foot and a foot, etc.[19] The inequality of numbers is given by the fact that when comparing numerical values, one number is greater or

---

12   *De arith. Boeth.*, pp. 126–150; see also Evans, "Introductions to Boethius's 'Arithmetica'…," p. 23.
13   See Caiazzo, "Un commento altomedievale…," pp. 123–125.
14   Boethius, *Arith.* II, 1, pp. 93–94; cf. Nicomachus, *Arith.* II, 1, 1–2, pp. 73–74.
15   Boethius, *Arith.* I, 2, p. 14; cf. Nicomachus, *Arith.* I, 6, 1, p. 12.
16   Cf., e.g., Boethius, *Trin.* 5, p. 28.
17   Boethius *Arith.* I, 32, p. 80; cf. Nicomachus, *Arith.* I, 23, 4, p. 65; see also Abbo, *In Calc.* III, 22–23, pp. 85–86.
18   Boethius *Arith.* I, 21, p. 54–55; cf. Nicomachus, *Arith.* I, 17, 2, p. 44.
19   Boethius, *Arith.* I, 21, p. 55; cf. Nicomachus, *Arith.* I, 17, 3, p. 44; see also Isidore, *Etym.* III, 6, 3; Cassiodore, *Inst.* II, 5, p. 136; or *De arith. Boeth.* ad I, I, p. 131.

lesser than the other – through this, the two basic types of inequality are also determined. Greater or lesser inequality is further divided into five basic types of inequality,[20] i.e., multiples and submultiples (i.e., divisors),[21] superparticular and subsuperparticular ratios,[22] superpartient and subsuperpartient ratios,[23] superparticular multiples and subsuperparticular divisors,[24] and, finally, superpartient multiples and subsuperpartient divisors.[25] For the schematic order, see Table 1.

---

20 Boethius, *Arith.* I, 22, p. 56; cf. Nicomachus, *Arith.* I, 17, 7–8, p. 45–6; see also Cassiodore, *Inst.*, II, 5, p. 135.
21 Boethius, *Arith.* I, 23, pp. 56–59; cf. Nicomachus, *Arith.* I, 18, pp. 46–8; see also Isidore, *Etym.*, III, 6, 5–6; Cassiodore, *Inst.* II, 5, p. 136; or Abbo, *In Calc.* III, 73, p. 117.
22 Boethius, *Arith.* I, 24, pp. 60–63; cf. Nicomachus, *Arith.* I, 19, pp. 49–55; see also Isidore, *Etym.* III, 6, 7 and III, 6, 10; Cassiodore, *Inst.* II, 5, pp. 136–137; Martianus, *De nupt.* VII, 761, pp. 279–280; or Abbo, *In Calc.* III, 74, pp. 117–118.
23 Boethius, *Arith.* I, 28, pp. 70–73; cf. Nicomachus, *Arith.* I, 20–21, pp. 55–59; see also Isidore, *Etym.* III, 6, 8–9; Cassiodore, *Inst.* II, 5, p. 137; Martianus, *De nupt.* VII, 762, p. 280; or Abbo, *In Calc.* III, 75–76, pp. 118–119.
24 Boethius, *Arith.* I, 29, pp. 73–78; cf. Nicomachus, *Arith.* I, 22, pp. 59–63; see also Isidore, *Etym.* III, 6, 11–13; Cassiodore, *Inst.* II, 5, pp. 137–8; or Abbo, *In Calc.* III, 77–78, pp. 119–120.
25 Boethius, *Arith.* I, 30, pp. 78–79; cf. Nicomachus, *Arith.* I, 23, 1–3, pp. 63–64; see also Isidore, *Etym* III, 6, 12–13; Cassiodore, *Inst.* II, 5, p. 138; or Abbo, *In Calc.* III, 79, pp. 120–121.

Table 1 – Ten types of inequality

				2 : 1	2, 4, 8
1.	multiple (*multiplex*)	a greater number contains a lesser number more than once with no remainder	$b \cdot a / a$; if $b > 1$ and $a \geq 1$	3 : 1	3, 6, 9
				4 : 1 etc.	4, 8, 12
				3 : 2	4, 6, 9
2.	superparticular ratio (*superparticularis*)	a greater number contains a whole lesser number exactly once plus one part of the lesser number expressed by a fraction whose numerator is number one	$(a+1)/a$; if $a > 1$	4 : 3	9, 12, 16
				5 : 4 etc.	16, 20, 25
				5 : 3	9, 15, 25
3.	superpartient ratio (*superpartiens*)	a greater number contains a whole lesser number exactly once plus more than one part of the lesser number, i.e., the part of the lesser number cannot be expressed by a fraction whose numerator would be number one	$(a+n)/a$; if $a > n > 1$	7 : 4	16, 28, 49
				9 : 5 etc.	25, 45, 81
				5 : 2	4, 10, 25
4.	multiple superparticular (*multiplex superparticularis*)	a greater number contains a lesser number more than once plus one part of a lesser number expressed by a fraction whose numerator is number one	$(b \cdot a + 1)/a$; if $b > 1$ and $a > 1$	7 : 2	4, 14, 49
				7 : 3 etc.	9, 21, 49
				8 : 3	9, 24, 64
5.	multiple superpartient (*multiplex superpartiens*)	a greater number contains a lesser number more than once plus more than one part of the lesser number, i.e., the part of the lesser number cannot be expressed by a fraction whose numerator would be number one	$(b \cdot a + n)/a$; if $b > 1$ and $a > n > 1$	11 : 4	16, 44, 121
				11 : 3 etc.	9, 33, 121

ONE NUMBER IS GREATER THAN THE OTHER

(continued on next page)

Table 1 Continued

				1:2	8, 4, 2
6.	divisor/submultiple (*submultiplex*)	a lesser number is contained within a greater number more than once without a remainder (the opposite ratio to a multiple)	a/b; if $a \geq 1$ and $b > 1$	1:3	9, 6, 3
				1:4 etc.	12, 8, 4
				2:3	9, 6, 4
7.	subsuperparticular ratio (*subsuperparticularis*)	a lesser number is contained exactly once within a greater number along with one other part of itself (the opposite ratio to a superparticular)	$a/(a+1)$; if $a > 1$	3:4	16, 12, 9
				4:5	25, 20, 16
				3:5;	25, 15, 9
8.	superpartient ratio (*subsuperpartiens*)	a lesser number is contained exactly once within a greater number along with more than one of its own parts (the opposite ratio to a superpartient)	$a/(a+n)$; if $a > n > 1$	4:7	49, 28, 16
				5:9 etc.	81, 45, 25
				2:5	25, 10, 4
9.	submultiple superparticular (*submultiplex superparticularis*)	a lesser number is contained within a greater number more than once along with one other part of itself (the opposite ratio to a multiple superparticular)	$a/(b \cdot a + 1)$; if $b > 1$ and $a > 1$	2:7	49, 14, 4
				3:7 etc.	49, 21, 9
				3:8	64, 24, 9
10.	submultiple superpartient (*submultiplex superpartines*)	a lesser number is contained within a greater number more than once along with more than one of its own parts (the opposite ratio to a multiple superpartient)	$a/(b \cdot a + n)$; if $b > 1$ and $a > n > 1$	4:9	121, 44, 16
				3:11 etc.	121, 33, 9

ONE NUMBER IS LESSER THAN THE OTHER

All these inequalities arise from the equality of the procedure, which Boethius demonstrates on examples of a three-member numerical sequence (see Table 2). If the three identical options are given, multiples arise first – initially doubles, then triples, quadruples, etc. Multiples are the basis for superparticular ratios that arise from reverse multiple sequences. Doubles give rise to a 3 : 2 ratio, triples to a 4 : 3 ratio, quadruples to a 5 : 4 ratio, etc. After re-reversal of a sequence arranged in superparticular ratio, a superpartient ratio arise: a 3 : 2 ratio is the root of a 5 : 3 ratio, a 4 : 3 ratio produces a 7 : 4 ratio, a 5 : 4 ratio establishes a 9 : 5 ratio, etc. Standardly (not reversibly) arranged superparticular ratios help us to gain superparticular multiples: from a 3 : 2 ratio arises a 5 : 2 ratio, from a 4 : 3 ratio arises a 7 : 3 ratio, and from a 5 : 4 ratio arises a 9 : 4 ratio. Similarly, according to the upwardly arranged superpartient ratios, superpartient multiples are created: a 5 : 3 ratio becomes a 8 : 3 ratio, a 7 : 4 ratio becomes an 11 : 4 ratio, and a 9 : 5 ratio becomes a 14 : 5 ratio.

Table 2 – The emergence of inequalities from equality

aequalitas	1 – 1 – 1 (ratio 1 : 1) ↓					
multiplex	1 – 2 – 4 (ratio 2 : 1) →		1 – 3 – 9 (ratio 3 : 1) →		1 – 4 – 16 (ratio 4 : 1) → etc.	
	conversus ↓		*conversus* ↓		*conversus* ↓	
superparticularis	4 – 6 – 9 (ratio 3 : 2)		9 – 12 – 16 (ratio 4 : 3)		16 – 20 – 25 (ratio 5 : 4)	
	conversus ↓		*conversus* ↓		*conversus* ↓	
superpartiens	9 – 15 – 25 (ratio 5 : 3)	↓	16 – 28 – 49 (ratio 7 : 4)	↓	25 – 45 – 81 (ratio 9 : 5)	↓
multiplex superparticularis	↓	4 – 10 – 25 (ratio 5 : 2)	↓	9 – 21 – 49 (ratio 7 : 3)	↓	16 – 36 – 81 (ratio 9 : 4)
multiplex superpartiens	9 – 24 – 64 (ratio 8 : 3)		16 – 44 – 121 (ratio 11 : 4)		25 – 70 – 196 (ratio 14 : 5)	

To understand the origin of these sequences in the given ratios, we need to know three simple rules. When converting a three-member sequence (i.e., from $a - b - c$ to $a_1 - b_1 - c_1$) from one ratio to another (e.g., to equality, i.e., 1 : 1 ratio; to multiples, i.e., 2 : 1 ratio) according to Boethius (and, of course, Nicomachus) the following applies:[26]

---

26 Boethius, *Arith.* I, 32, p. 81; cf. Nicomachus, *Arith.* I, 23, 8, p. 66.

[P1] $a_1 = a$
[P2] $b_1 = a + b$
[P3] $c_1 = a + 2b + c$

By applying the three rules, we can reconstruct the emergence of all kinds of inequalities from original identity and unity, which is the dominant framework within which especially Abbo of Fleury discusses these issues. Like Boethius, he claimed that every philosopher must address the disciplines of the *quadrivium* from a philosophical point of view, since without knowledge of these four arts, one cannot arrive at the truth. Abbo assumed that philosophy, the love of wisdom, is also the love of God.[27] The knowledge of arithmetic leads us safely to God, for God's wisdom (the symbol of which is the Temple of Wisdom built on seven pillars, corresponding to the seven liberal arts[28]) created everything according to its measure, number, and weight.[29] Consequently, in the beginning, there could not have existed anything other than a unit, which is the principle and origin of all numbers: all numbers are from the unit, are through the unit, and are in the unit.[30] In addition, as it has no parts, it is therefore identical to being and necessity,[31] and thus it is the divine foundation of the universe, which creates everything else according to mathematical ratios – the so-called *lambda diagram* in relation to Plato's *Timaeus*, and Calcidius's and Macrobius's commentaries on it.[32]

The author of the commentary called *De arithmetica Boetii* sees the importance of the knowledge of transfers between sequences in a very similar light: everything is either the Creator or the Created. Everything created comes from the Creator, just as inequality hinges on original equality.[33] Equality – and the Highest Good – is identical to God and only inequality and evil arise if we move away from this original One.[34] The knowledge of the return to equality is the knowledge of the return to God, the primary Good and equality.

However, the debate over Boethius's *Arithmetic* prior to the year 1000 was provoked not so much by the last chapter of the first book, as by the opening

---

27 Abbo, *In Calc.* II, 1, p. 65 or II, 7, p. 68.
28 Ibid. II, 1, p. 65; cf. *Prov.* 9, 1.
29 See, for example, Abbo, *In Calc.* II, 10–11, p. 69; cf. *Sap.* 11, 20
30 Abbo, *In Calc.* III, 5, p. 75; cf. *Rom.* 11, 36.
31 Abbo, *In Calc.* III, 3, p. 74.
32 Ibid. III, 1–2, pp. 72–4; cf. Plato, *Tim.* 34b–36d; Macrobius, *In Somn.* I, 6, 46, p. 26; Calcidius, *In Tim.* I, 32, pp. 81–82.
33 *De arith. Boeth.* ad I, XXXII, pp. 140–141.
34 Ibid., pp. 141–142.

chapter of the second book. After explaining the birth of multiplicity in original unity, Boethius addresses the issue of returning inequality to original equality. If three numerical values are given, the organization of which corresponds to a certain ratio, then regardless of the type of inequality (multiples, superparticular or superpartient ratios, superparticular, or superpartient multiples), they can be converted to equality.[35]

In this case also, three rules are given, which must be followed for a valid search for the return to initial equality:[36]

[R1] $a_1 = a$
[R2] $b_1 = b - a$
[R3] $c_1 = c - (2b_1 + a)$

Immediately after the presentation of these rules, a passage follows which was probably the decisive impulse for a unique debate among early medieval thinkers: Boethius further states that by the application of these rules, we are able to convert a sequence of three numerical values ordered in a certain proportion to a more fundamental ratio. With regard to multiples, the following rule applies: quadruples are converted to triples, and triples to doubles that are then converted back to unity. Regarding superparticular ratios, a 5 : 4 ratio is changed to a more authentic 4 : 3 ratio; a 4 : 3 ratio subsequently gives rise to a 3 : 2 ratio, from which three identical numerical values are derived.[37] Boethius provides only one example to support this explanation, in which quadruples gradually return to equality[38] – see Table 3.

Table 3 – Boethius's example of converting multiples to equality according to [R1–3] rules

1.	8 – 32 – 128	quadruples (4 : 1 ratio)	the application of rules *[R1–3]*	↓
2.	8 – 24 – 72	triples (3 : 1 ratio)	the application of rules *[R1–3]*	↓
3.	8 – 16 – 32	doubles (2 : 1 ratio)	the application of rules *[R1–3]*	↓
4.	8 – 8 – 8	equality (1 : 1 ratio)		

---

35   Boethius, *Arith*. II, 1, p. 94; cf. Nicomachus, *Arith*. II, 1, 2, p. 74.
36   Boethius, *Arith*. II, 1, p. 94; cf. Nicomachus, *Arith*. II, 2, 1, pp. 74–75.
37   Boethius, *Arith*. II, 1, p. 94.
38   Ibid. II, 1, pp. 95–96.

After this example Boethius suggests that everyone who follows the presented rules will easily be able to find harmony (*convenientia*) in inequality, since identity is the mother of all inequalities: all inequality issues from it and all inequality returns to it.[39]

Despite this detailed analysis of the issues, one relatively important question remains unresolved: how exactly do we transfer each type of inequality? Boethius considers this part of arithmetic the deepest (*profundissima*) science,[40] and the author of the commentary *De arithmetica Boetii* adds that it is the most mysterious and subtle (*obscurissima et subtilissima*) of teachings.[41] However, uncertainties remain, the resolution of which is not apparent. If we limit ourselves here (in accordance with the main differences in approaches to the issue that emerged among the authors of the last quarter of the 10th century) upon the mutual relationship of superparticular ratios, multiples, and equality, then an alternative reading of Boethius's text is possible:

1. In accordance with *Introduction to Arithmetic* II, 1 it is possible to convert different types of superparticular ratios (5 : 4, 4 : 3, 3 : 2) and multiples (4 : 1, 3 : 1, 2 : 1) to equality (1 : 1) within the ranks of one type of inequality, i.e., for example, from triples to doubles and then to identity, respectively, from a 4 : 3 ratio to a 3 : 2 ratio, and from these ratios to the same numerical values.
2. If the process of the rise of inequality from equality and the return of inequality to equality is reversible, which follows from Boethius's text, it is not possible (except for multiples) to always act exclusively within the given inequality during the search for original equality, since a 3 : 2 ratios do not have their origin in equality but in doubles, a 4 : 3 ratios in triples, etc., according to *Introduction to Arithmetic* I, 32.

In other words, if the rules from *Introduction to Arithmetic* II, 1 are followed, one cannot comply with the instructions from *Introduction to Arithmetic* I, 32, and vice versa. It seems that this very alternative interpretation of the two chapters from Boethius's treatise was key to the whole debate on the conversion of ratio sequences at the end of the 10th century.

Notker of Liège inclined towards the first option. Following Boethius, he attempts to show that superparticular ratios can be converted among themselves

---

39 Ibid., p. 96.
40 Ibid. I, 32, p. 80.
41 *De arith. Boeth.* ad I, XXXII, p. 140; similarly, Gerbert, *In Boeth. Arith.* 1, p. 32.

and only during the penultimate step, when a 3 : 2 ratio is converted to a double, is it possible to proceed directly to equality.

Notker, however, was well aware that the reduction of superparticular relationships between themselves, without immediate conversion to multiples, put him at odds with the three rules for transitions between different ratios (i.e., rules [R1-3]) emphasised so keenly by Boethius. Perhaps due to this, the brief text begins with the caveat that Boethius did not intend his rule to be the only possible method for transferring three unequal values arranged in a certain ratio to initial equality – it was merely an auxiliary tool (*adjuvabit*) that we do not need to apply unconditionally in all cases. According to Notker, it is definitely not suitable when looking for a path from the three numerical values of a 5 : 4 relative sequence to the three identical items.[42]

Notker proposes different rules for how to proceed while making individual conversions. In the case of transition from a 5 : 4 to a 4 : 3 ratio, he recommends the following:[43]

[N1] $a_1 = a$
[N2] $b_1 = b - 1/2a$
[N3] $c_1 = c - a$

For transition between a 4 : 3 and a 3 : 2 ratio he suggests observing these rules:[44]

[O1] $a_1 = a$
[O2] $b_1 = b - 1/2b$; (or $b_1 = 1/2b$)
[O3] $c_1 = c - b$

The search for original equality of three values within a sequence arranged in a superparticular way appears, according to Notker, as follows: a 5 : 4 ratio transfers to a 4 : 3 ratio, then to a 3 : 2 ratio, which converts to double, and, subsequently, acquires equality, as presented in Table 4.[45]

---

42 Notker, *Superpart.*, p. 297.
43 Ibid., p. 298.
44 Ibid.
45 Ibid.

Table 4 – Notker's approach for conversion of a 5 : 4 ratio to equality

1.	A) 5 : 4 ratio	16 – 20 – 25	the application of rules [N1–3]	↓
2.	[3 : 4 ratio]	16 – 12 – 9	the reverse of sequences	↓
3.	B) 4 : 3 ratio	9 – 12 – 16	the application of rules [O1–3]	↓
4.	[2 : 3 ratio]	9 – 6 – 4	the reverse of sequences	↓
5.	C) 3 : 2 ratio	4 – 6 – 9	the application of rules [R1–3]	↓
6.	[1 : 2 ratio]	4 – 2 – 1	the reverse of sequences	↓
7.	D) double (2 : 1)	1 – 2 – 4	the application of rules [R1–3]	↓
8.	E) equality (1 : 1)	1 – 1 – 1		

Notker's process of transition between individual superparticular ratios does not follow the fundamental steps of Boethius's rules [R1–3] for the gradual reduction and return, but it is undoubtedly the correct approach and, moreover, agrees with Boethius's explicit thesis that superparticular ratios can be transferred without intermediate ratios. However, this is not how the text of *De superparticularibus* to Boethius's *Introduction to Arithmetic* ends. The author continues with the way individual inequalities are created, and shows how three kinds of inequality can be converted to superparticular ratios, from which it is clear how to arrive at equality.

Superpartient ratios are converted to superparticular according to the following rules:

[S1 = R1] $a_1 = a$
[S2 = R2] $b_1 = b - a$
[S3] $c_1 = c - [x + (x - y) / y]a$; or $c_1 = c - (2x - y / y)a$) where x is the numerator of the given ratio (the first number) and y is the denominator (the second number) of the given ratio.[46]

---

46   Notker in his brief text does not give the [S3] rule in the form just presented, but for each case of superpartient ratio (x / y) he refers to a specific value of the first number (a) of the given ratio that must be subtracted from the third number (c) to get the third entry of a new sequence (c_1). So, for example, for a 9 : 5 ratio it is necessary to subtract the first entry of a numerical sequence twice from a third entry and even three-fifths of the first item (i.e., 13 / 5a), which corresponds with [S3] the rule for a ratio of 9 : 5, then $x = 9$, $y = 5$, that is ($x + (x - y) / y$) and $(9 + (9 - 5) / 5) a = 13 / 5a$. In the case of a 7 : 4 ratio, according to the [S3] rule, the number c subtracts the number a twice and then a half of the number (i.e., 5 / 2a), since $x + (x - y) / y$) and $(7 + (7 - 4) / 4)a = 10 / 4a = 5 / 2a$. It is same for 5 : 3 ratio, where the [S3] rule urges us to subtract from an item c twice item a and even the third of item a (i.e. 7 / 3a), since from ($x + (x - y) / y$) a arises (5 + (5 - 3) / 3)$a = 7 / 3a$. See Notker, *Superpart.*, pp. 298–299: in the text the value of 48 is mistakenly listed instead of the correct 49 concerning the third item of a 7 : 4 ratio.

The Nature of Numbers: *Saltus Gerberti* (Letter to Constantine) 47

Notker then moves on to the fourth and fifth kind of inequality, i.e., superparticular and superpartient multiples. Here he gives only one example, which, moreover, remains incomplete for superpartient multiples. However, it is relatively easy to formulate the basic rules by which one should proceed in these cases:

[T1 = R1] $a_1 = a$
[T2 = R2] $b_1 = b - a$
[T3] $c_1 = c - [(x / y) a + b]$, or $c_1 = c - [(z - ny) / y]a + b$, where x / y is superparticular or superpartient ratio to which a particular sequence is transferred; z is the numerator of a superpartient or superparticular multiple, which is subtracted from the multiples reduced by one (i.e., n); y is the denominator of a superparticular or superpartient multiple.

Converting a superparticular multiple results in a superparticular ratio, from which it is already possible to reach a double and, thus, equality, as has been previously shown. When a superpartient multiple is converted, firstly, a superpartient ratio arises, which it is necessary to transform to the superparticular ratio before proceeding in the same way. Using the rules [T1–3] is relatively easy. If it is necessary to reduce the multiples of two and one third (i.e., a 7 : 3 ratio), then $x / y = 4/3$, since the rise of superparticular 4 : 3 ratio, or $(x - ny) / y = (7 - 1 \cdot 3) / 3 = 4/3$, then according to [T3] rule, the third item of a derived sequence (c_1) is given by the difference of the c item and the sum of $4/3a$ with b.[47] Similarly, in the case of a superpartient multiple, it is clear that, for example, in the multiple of two and three-quarters, i.e., 11 : 4 ratio, then $x / y = 7/4$, respectively, $(x - ny) / y = (11 - 1 \cdot 4) / 4 = 7/4$, the third item of the numerical sequence needs to be transferred to a superpartient ratio (7 : 4) by applying the [T3] rules, i.e., $c_1 = c - 7/4a + b$.[48]

Through this, Notker has discussed the conversions of all the inequalities to their predecessors which ultimately lead to equality. Since the only surviving manuscript of the text ends in mid-sentence, we have no choice but to reconstruct the return of individual inequalities to original equality without direct reference to Notker's *scholium*. It appears, however, that the returns of superparticular and superpartient multiples to equality might resemble what is shown in Tables 5 and 6.

---

47 Notker, *Superpart.*, pp. 299. The sequence of a multiple of two and one-third is given in the text with two erroneous values: the second item should be 21 instead of 15, which is given, and the correct third value is 49, not 48.
48 Notker, *Superpart.*, p. 299.

Table 5 – The conversion of superparticular multiples to equality according to Notker

1.	A) superparticular multiple	9 : 4	16 – 36 – 81	the application of rules [T1–3]	↓
2.	B) superparticular ratios	5 : 4	16 – 20 – 25	the application of rules [N1–3]	↓
3.			16 – 12 – 9	the reverse of sequences	↓
4.		4 : 3	9 – 12 – 16	the application of rules [O1–3]	↓
5.			9 – 6 – 4	the reverse of sequences	↓
6.		3 : 2	4 – 6 – 9	the application of rules [R1–3]	↓
7.			4 – 2 – 1	the reverse of sequences	↓
8.	C) multiple	2 : 1	1 – 2 – 4	the application of rules [R1–3]	↓
9.	D) equality	1 : 1	1 – 1 – 1		

Table 6 – The conversion of superpartient multiples to equality according to Notker

1.	A) superpartient multiple	11 : 4	16 – 44 – 121	the application of rules [T1–3]	↓
2.	B) superpartient ratio	7 : 4	16 – 28 – 49	the application of rules [S1–3]	↓
3.			16 – 12 – 9	the reverse of sequences	↓
4.	C) superparticular ratio	4 : 3	9 – 12 – 16	the application of rules [O1–3]	↓
5.			9 – 6 – 4	the reverse of sequences	↓
6.		3 : 2	4 – 6 – 9	the application of rules [R1–3]	↓
7.			4 – 2 – 1	the reverse of sequences	↓
8.	D) multiple	2 : 1	1 – 2 – 4	the application of rules [R1–3]	↓
9.	E) equality	1 : 1	1 – 1 – 1		

Although it may not be obvious at first sight, Notker's rules [S3] and [T3] provide the same reduction of sequences as Boethius's rule [R3], since all the rules are mutually convertible. Notker created modified rules to enable transitions from all types of inequality. However, they do not follow the wording of Boethius's rules, and also, in the case of superparticular multiples, do not proceed in reverse the order of the derivation of inequality. These were probably the main reasons why Abbo of Fleury and Gerbert of Aurillac objected to these instructions.

While Notker chose the first reading of Boethius's *Introduction to Arithmetic*, Gerbert and Abbo instead chose the second (i.e., holding to [R1–3] rules) and interpreted Boethius's statement that the individual superparticular ratios should be transferred between each other differently.

Gerbert wrote very precise and detailed instructions for the transition from superparticular 5 : 4 ratios to quadruples (4 : 1), and from quadruples to triples (3 : 1), which are also the means for the proper transition to superparticular 4 : 3 ratios. This method of converting a 5 : 4 ratio to a 4 : 3 ratio, with direct reference to Boethius, was introduced as correct.[49] A 4 : 3 ratio can revert to triples (3 : 1) to provide doubles (2 : 1) in the next step and, in accordance with the emergence of inequality, it applies that doubles give rise to a 3 : 2 ratio, which concludes the path from 5 : 4 ratios to 3 : 2 ratios according to Boethius's rules, and avoids any possible confusion.[50] The use of Boethius's reduction rules will readily provide equality (1 : 1) from doubles, and Gerbert notes for the fourth time in his short text that this is the only appropriate method (i.e., in accordance with Boethius and, in particular, in accordance with the nature of numbers, numerical relationships, and their origin) of converting superparticular ratios to lower ratios of the same inequality and, if necessary, to equality itself.[51] The sequence of *Saltus Gerberti* (Gerbert's Leap) is offered in Table 7.

Table 7 – The so-called *Saltus Gerberti*, i.e. the transition between superparticular ratios and multiples

1.	sesquiquartum	5 : 4	16 – 20 – 25	the application of rules [R1–3]	↓
2.	quarter	1 : 4	16 – 4 – 1	the reverse of sequences	↓
3.	quadruple	4 : 1	1 – 4 – 16	the application of rules [R1–3]	↓
4.	triple	3 : 1	1 – 3 – 9	the reverse of sequences	↓
5.	third	1 : 3	9 – 3 – 1	the application of rules [P1–3]	↓
6.	sesquitertium	4 : 3	9 – 12 – 16	the application of rules [R1–3]	↓
7.	third	1 : 3	9 – 3 – 1	the reverse of sequences	↓
8.	triple	3 : 1	1 – 3 – 9	the application of rules [R1–3]	↓
9.	double	2 : 1	1 – 2 – 4	the reverse of sequences	↓
10.	half	1 : 2	4 – 2 – 1	the application of rules [P1–3]	↓
11.	sesquialterum	3 : 2	4 – 6 – 9	the application of rules [R1–3]	↓
12.	half	1 : 2	4 – 2 – 1	the reverse of sequences	↓
13.	double	2 : 1	1 – 2 – 4	the application of rules [R1–3]	↓
14.	equality	1 : 1	1 – 1 – 1		

Gerbert's letter seems, quite clearly, to be a reaction to Notker's (or a similar) reading of Boethius. The repeated refusal of immediate transfers of superparticular ratios and the repeated recommendation to always honour Boethius's rules [R1–3] or [P1–3], since only through their usage can we follow

---

49   Gerbert, *In Boeth. Arith.* 1, pp. 32–34.
50   Ibid. 2, pp. 34–35.
51   Ibid. 2, p. 35.

the process of how the numerical sequences were created, seems to be transparently targeted at the different procedure introduced in Notker's text. Gerbert did not think that Boethius, with his claims about the transfer of a 5 : 4 ratio to a 4 : 3 ratio, etc., meant an immediate transition from one superparticular ratio to another, but assumed intermediate stages in the form of multiple inequalities.

Abbo of Fleury addresses the given problem of Boethius's *Introduction to Arithmetic* in the same way as Gerbert. Directly quoting Boethius's problematic statement about the mutual transfers of superparticular ratios, he immediately interprets it – Boethius meant that between a 4 : 3 and a 5 : 4 ratio there can never be another superparticular ratio, but we cannot take it that these ratios transfer each other without any intermediary, as superparticular ratios originate from multiples and, therefore, must always be converted to multiples.[52]

Consequently, it is necessary to honour Boethius's rules [**R1-3**],[53] and in accordance with the reverse order of dissimilar numerical sequences, come to original equality – i.e., from superpartient multiples to superpartient ratios, from these ratios to superparticulars, and through multiples to equality.[54] Abbo subsequently presents two tables showing how the individual inequalities appeared and how we can return to perfect unity.[55]

Gerbert and Abbo deal with the consequences arising from *Introduction to Arithmetic* II, 1 in identical fashion and are content to observe the emergence of inequality (according to *Introduction to Arithmetic* I, 32) which it is necessary to revert to during a return to embryonic unity and equality, after which we proceed according to the rules [**R1-3**]. Notker's instructions for implementing the transition are possible, but they reveal signs of confusion. In particular, they are not in accordance with the nature of numbers and the emergence of proportional sequences.

As the previous paragraphs suggest, the theme of the introductory chapter of Boethius's *Introduction to Arithmetic* attracted considerable attention in the late 10th century, which automatically raises the question of why there was such interest in this problem – seemingly, at first sight, a very specific, and purely mathematical phenomenon. However, a metaphysical-creationist element had already been introduced and undoubtedly formed the framework defining the whole debate: i.e., how could anyone consider themselves a thorough believer

---

52   Abbo, *In Calc.* III, 23, p. 87 and III, 27, p. 89; cf. Boethius, *Arith.* II, 1, p. 94.
53   Abbo, *In Calc.* III, 22, p. 86.
54   Ibid. III, 23–24, p. 87.
55   Ibid. III, 25–26, p. 88.

if they did not strive to reach God? If mathematical truth, numbers, numerical ratios, etc., are instruments that God used in the creation of the universe, when he allowed the emergence of diversity derived from the fundamental unity and identity, and if everything is transferred back in the same order to God in the end, then it is clear that not only Neo-Pythagoreans and Pythagoreans, not only Plato and Neoplatonists but also Christians are almost obliged to explore this mysterious secret of creation and the return to the Creator. However, the metaphysical-theological and anthropological context that shaped interest in the detailed analysis of the mentioned sequences is far from the only one in which the issue of numerical ratios (and their resulting numerical series) can be included.

Questions related to sequences arranged by numerical ratios are not explored only by arithmetic but also by music, according to the definition of the subjects of the individual *quadrivium* disciplines, as already mentioned in the introduction. The theory of the art of music very intensively deals with ratios, their mutual relations, connections, transfers, etc. These are topics that can be historically traced back to the times of Pythagoras of Samos himself – for example, the famous story about finding ratios in a blacksmith's workshop, which Boethius recounts.[56]

The application of numerical ratios to music theory (e.g., chord tones, monochords, etc.) were, and are, fully automatic, which of course concerned early medieval authors, who often understood this issue as a natural transition between arithmetic and musical arts, as expressed, for example, in the words of Abbo of Fleury.[57] Thematically, it is expressed also by the author of the text *De arithmetica Boetii*, where he, following Boethius, explains the identity of arithmetical and musical ratios: *diatessaron – sesquitertia* (4 : 3); *diapente – sesqualtera* (3 : 2); *diapason – dupla* (2 : 1) and *epogdous – sesquioctavus* (9 : 8).[58] Similarly, Gerbert of Aurillac also commented on Boethius's *Introduction to Music*, as mentioned above, taking an intense interest in the organ, and using a monochord.[59]

---

56 Boethius, *Mus.* I, 10, pp. 196–198.
57 Abbo, *In Calc.* III, 83, pp. 122–123.
58 *De arith. Boeth.*, I, I, p. 134 and ad II, II, p. 144; cf. also Boethius, *Arith.* II, 48, pp. 200–201, idem, *Mus.* I, 16–19, pp. 201–205; Abbo, *In Calc.* III, 2, p. 74; etc.
59 See, e.g., Gerbert, *2 In Boeth. Mus.*, pp. 30–31; idem, *Ep.* 92, p. 121–122; or idem, *Rogatus*, pp. 59–81; cf. also Sachs, K. J., "Gerbertus cognomento musicus. Zur musikgeschichtlichen Stellung des Gerbert von Reims (nachmaligen Papstes Silvester II)." *Archiv für Musikwissenschaft* 29 (1972), pp. 257–274; or Richer, *Hist.* III, 49, p. 195.

However, music is not the only liberal art in which the arithmetical principles of these sequences are actively used. This is illustrated by the initial passage of the Munich manuscript from the monastery of Tegernsee CLM 18764, fol. 1v, in which numerical ratios are used to interpret metrical feet, traditionally an area of grammar. All the feet are divided according to the proportion in which the stressed and unstressed sections (so-called long and short syllables) are arranged. The author, following Isidore's *Etymologiae*[60] distinguishes:[61]

1. an equal ratio: i.e., the numerical ratio of one foot is composed of equal parts, i.e., it represents equality – a 1 : 1 ratio, such as the numbers 2 and 2, or 4 and 4;
2. a double: i.e., the ratio within one foot is such that the smaller number is twice exceeded by the larger, and hence the larger number contains in itself the smaller number twice, i.e., a 2 : 1 ratio, such as the numbers 1 and 2, or 2 and 4;
3. a sesquialterum: i.e., this foot is designed with a ratio in which the larger number includes the smaller number and half of it again, (the smaller number is contained in the larger number one and a half times, i.e., a 3 : 2 ratio);
4. a triple: i.e., a foot in which the larger number exceeds the smaller three times, i.e., a 3 : 1 ratio;
5. a sesquitertium: i.e., a foot in which the larger number contains the whole smaller number and a third of the smaller number again, i.e., a 4 : 3 ratio.

In addition to this classification, there is also a list of individual metric feet: ten with the same parts – pyrrich, spondee, dactyl, anapest, proceleumaticus, dispondee, diiamb, ditrochee, antispast and choriamb; six with double portions – iamb, trochee, molossus, tribrach and major or minor ionic; only one metric foot with a triple – amphibrach; seven feet contain sesquialterum (3 : 2) multiples – cretic or amphimacer, bacchius, antibacchius and first to fourth paeon; and four contain sesquitertium (4 : 3) multiple parts – from the first to the fourth epitrite.[62] This classification obviously represents another significant area that actively uses numerical ratios and their relationships.

However, the list of the liberal arts which actively used numerical sequences is still not complete; yet another area can be mentioned to which numerical ratios and the ability to mutually transfer them belong as an inherent property – intellectual

---

60   Isidore, *Etym.* I, 17, 21–27.
61   Evans, "Introductions to Boethius's 'Arithmetica'...", p. 23.
62   Ibid.

recreation, i.e., games. It seems most likely that it was around the turn of the first millennium (thus at the same time as the arithmetical texts which this chapter deals with first appeared) that people in the Christian Latin West started to play the intellectual mathematical board game *rithmomachia*, i.e., the battle of the numbers. The oldest known texts about it (the so-called *Regula de rithmomachia, id est de numeri pugna* and *Ludus qui dicitur rithmimachia*) date from the first half of the 11th century.[63]

In the Middle Ages, the setting of the rules of the game was in the Middle Ages attributed to Gerbert of Aurillac (together with, for example, Boethius and Pythagoras);[64] however, it is more likely to have been formulated by someone from his immediate circle or by one of close to him or his epigones.[65] In manuscripts of the given era, tables on rithmomachia, auxiliary tools, or texts are commonly linked to arithmetical treatises. It is clear that the game itself is, in its earliest texts, always referred to as arithmetic,[66] and the issues addressed by Notker, Gerbert, and Abbo serve as necessary equipment for a player of rithmomachia. For this reason, rithmomachia is often regarded as a teaching aid for understanding Boethian theoretical arithmetic. Without a detailed presentation of the rules of the game or an illustration of the use of work with numerical ratios, it is sufficient if the starting position of the chips on the board is outlined. This in itself reflects Nicomachean-Boethian teaching about the relative properties of numbers and sequences (see Fig. 1).

---

63   For more detailed information of these two treatises on *rithmomachia* see Borst, A., *Das mittelalterliche Zahlenkampfspiel*. Heidelberg: Carl Winter, 1986, pp. 50–97.
64   Folkerts, M., " 'Rithmomachia', a Mathematical Game from the Middle Ages." In: idem, *Essays on Early Medieval Mathematics. The Latin Tradition*. Aldershot: Ashgate, 2003, p. XI-5.
65   See, for example, Moyer, A. E., *The Philosophers' Game. Rithmomachia in Medieval and Renaissance Europe*. Ann Arbor: University of Michigan Press, 2001, pp. 20–22; Silva, J. N., "Teaching and playing 1000 years ago, Rithmomachia." In: Sigismondi, C. (ed.), *Orbe novus. Astronomia e Studi Gerbertiani*. Roma: Universitalia, 2010, pp. 146–147; or idem, "Mathematical games in Europe around the year 1000." *GERBERTVS – International Academic Publication on History of Medieval Science* 1 (2010), p. 224, etc.
66   Cf. Asilo, *Rith*. 1, p. 330, or Hermann, *Rith*. 1, p. 335.

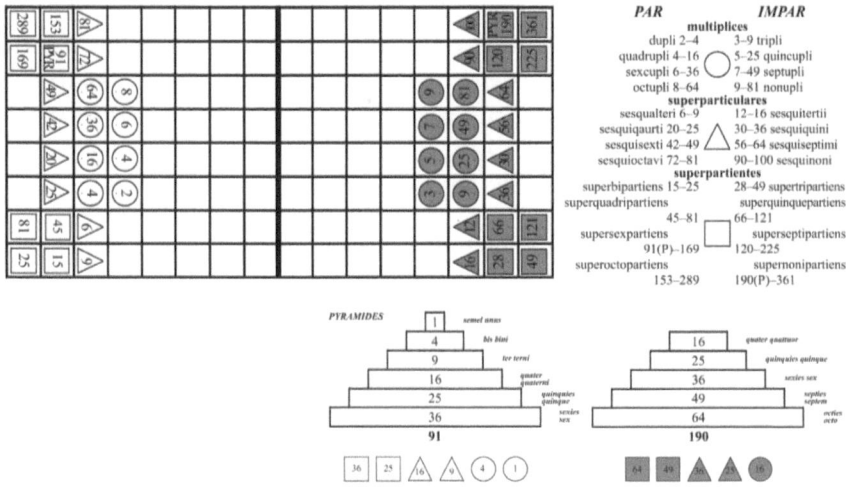

Fig. 1 – Rithmomachy, the starting position of chips

On one side of the board, some chips reflect even ratios, while on the other side, there are odd ratios. The values of the tokens reflect multiples, superparticular, and superpartient numbers.[67] Multiples of the even side are represented by a double (2 – 4), a quadruple (4 – 16), a sextuple (6 – 36), and an octuple (8 – 64); while multiples on the odd side are represented by a triple (3 – 9), quintuple (5 – 25), septuple (7 – 49), and nonuple (9 – 81). Superparticular numbers in the even group represent the following ratios: 3 : 2 (6 – 9), 5 : 4 (20 – 25), 7 : 6 (42 – 49), and 9 : 8 (72 – 81); and in the odd arrangement ratios such as 4 : 3 (12 – 16), 6 : 5 (30 – 36), 8 : 7 (56 – 64), and 10 : 9 (90 – 100). In both groups, four superpartient numbers are also present: the even enclave includes ratios 5 : 3 (15 – 25), 9 : 5 (45 – 81), 13 : 7 (91 – 169), and 17 : 9 (153 – 289), i.e., a larger number is composed of a smaller number and of two, four, six, or eight parts of a smaller number; the odd ratios set includes 7 : 4 (28 – 49), 11 : 6 (66 – 121), 15 : 8 (120 – 225), and 19 : 10 (190 – 361), i.e., a larger number is composed of a smaller number of three, five, seven, and nine parts of a smaller number.[68]

Both sides have a special figure, the so-called Pyramid, underlined in Fig. 1, and within the game itself, it is not made of one piece like other chips but of

---

67   Cf. Asilo, *Rith.* 2, pp. 330–331, or Hermann, *Rith.* 2, p. 335.
68   Cf. Asilo, *Rith.* 3–5, p. 331, or Hermann, *Rith.* 3, p. 336.

several chips put together. In the even team it represents a value of 91, which is made up of six pieces which correspond to the values of the first six square numbers: 1 (i.e., 1^2, 1 · 1), 4 (i.e., 2^2, 2 · 2), 9 (i.e., 3^2, 3 · 3), 16 (i.e., 4^2, 4 · 4), 25 (i.e., 5^2, 5 · 5), and 36 (i.e., 6^2, 6 · 6), wherein the sum of the values (1, 4, 9, 16, 25, and 36) gives 91. The odd-numbered side of the Pyramid consists of five chips and represents a value of 190, which is the sum of the representatives of the five square numbers: 16 (i.e., 4^2, 4 · 4), 25 (i.e., 5^2, 5 · 5), 36 (i.e., 6^2, 6 · 6), 49 (i.e., 7^2, 7 · 7), and 64 (i.e., 6^2, 8 · 8).[69]

Even from this starting position of game pieces with numbers on the game board, it becomes clear that without a knowledge of arithmetic, especially of working with numerical ratios, it would not be possible to play this game that was so popular in the Middle Ages. In addition, use of the game would fully correspond with Gerbert's often declared aim of the practical application of theoretical ideas. Finally, the so-called *Saltus Gerberti*, addressed to Constantine, could quite likely have served as a guide for the understanding of the basics of this intellectual game.

2. Figurative Numbers and Geometry (Letter to Adelbold)

As mentioned in the introduction, numbers were most often defined as collections of units during late antiquity and the early middle-ages. In his translation of Nicomachus' arithmetical treatise, Boethius stresses that numerical values can be expressed by various methods. Usually, when we want to write down, for example, the value '10', we use a numeral (Boethius would write the Roman numeral X; Nicomachus would naturally use the Greek numeral ι).[70] This method of numeral notification follows a certain custom (*usus*) and it does not reflect the natural (*naturalis*) composition of numbers. Since numbers are composed of units (*unitates*), their nature corresponds to graphical expressions using specific units, e.g., vertical lines (*virgulae*) or points (*puncti*), etc.[71] The number of units we write down corresponds to the natural value of the number; therefore, the number 'ten' may appear as follows: ||||||||||[72] (according to Nicomachus, it may look like this: ααααααααααα[73]).

---

69   Cf. Asilo, *Rith.* 8, pp. 332–333, or Hermann, *Rith.* 4, p. 337.
70   Boethius, *Arith.* II, 4, p. 106; Nicomachus, *Arith.* II, 6, 2, p. 83.
71   Boethius, *Arith.* II, 4, p. 107; Nicomachus, *Arith.* II, 6, 3, pp. 84–85.
72   Boethius, *Arith.* II, 4, p. 107.
73   Nicomachus, *Arith.* II, 6, 2, p. 84.

These units, which express the value of concrete numbers, might be sorted into various geometrical shapes. Through this, numbers attain specific figural characteristics, which are a traditional topic for numerical theories characteristic of arithmetic from at least the age of the Pythagoreans,[74] and which were also developed by Plato, for whom regular geometric bodies (so-called Platonic solids) served as a basis for geometrical interpretations of physical reality.[75] The fundamental shape is a triangle (*triangulum*), i.e., a plane number (*figura plana, superficies*), which forms the basis for almost all other plane numbers.[76] According to Plato, everything that comes into being consists of triangles, since three-dimensional objects have their origins in two dimensions and the source of the two-dimensional is a triangle.[77] Through this, plane (and, consequently, solid) figural numbers, as explored by arithmetic, form a basis for a metaphysical interpretation of everything we perceive as three-dimensional reality.

Basically, all plane numbers are derived from triangular numbers. Although we have numbers that are quadrangular, pentagonal, etc., the number of angles is always determined by the number of triangles located in the figural number – i.e., triangular numbers contain three triangles, quadrangular numbers contain four triangles, and pentagonal numbers contain five triangles, etc. (cf. Fig. 2).[78]

triangulum   quadratum   pentagonum   exagonum

Fig. 2 – Triangular basis of plane numbers

Thus, triangular numbers form the basis of other plane numbers and, furthermore, they are crucial to the philosophical interpretation of reality. Therefore, it is not surprising that triangular numbers are derived from natural numbers (i.e., positive integers). It is possible to extract the list of triangular numbers by

---

74 Cf., e.g., Deza, E. – Deza, M. M., *Figurate Numbers*. Singapore – Hackensack: World Scientific, 2012, pp. xv–xvii.
75 Plato, *Tim.* 55c–56c; cf. Euclid, *Elem.* XIII, 13–17, pp. 158–180.
76 Boethius, *Arith.* II, 6, p. 114; cf. Nicomachus, *Arith.* II, 7, 4, p. 87.
77 Plato, *Tim.* 53c–55c.
78 Cf. Gerbert [?], *Geom.* V, 1, pp. 71–72.

Figurative Numbers and Geometry (Letter to Adelbold)

forming a sequence of all natural numbers (starting with the mother of all numbers, i.e., 'one') and their successive adding up: at first we add up the first two values, to this sum we add the third value, to this sum the fourth value etc. The lowest triangular number is, thus, the sum of the unit and the first number (i.e., 'two'), which equals three – 'one' itself is only a potential triangle, rather than an actual triangle. The next triangular number is then 'six', followed by 'ten', etc. The process of their discovery is shown in Table 8; for their figural expression, see Fig. 3.

Table 8 – Forming the sequence of triangular numbers

Natural numbers (x_n)	(1)	2	3	4	5	6	7	8	etc.
$a_n = x_n + a_{n-1}$		2 + 1	3 + 3	4 + 6	5 + 10	6 + 15	7 + 21	8 + 28	etc.
Triangular numbers (a_n)	(1)	3	6	10	15	21	28	36	etc.

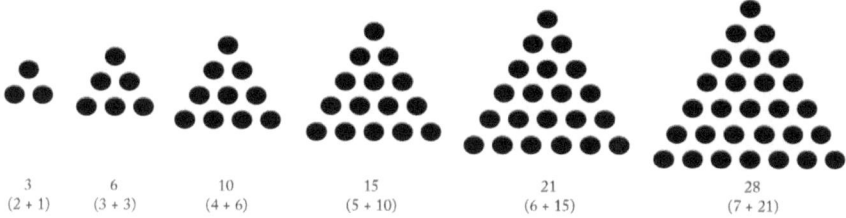

```
   3        6       10       15        21         28
 (2 + 1)  (3 + 3) (4 + 6)  (5 + 10)  (6 + 15)  (7 + 21)
```

Fig. 3 – Triangular numbers

In accordance with the statement that triangular numbers form the basis for other plane angular numbers, it is possible to derive their sequences mathematically from triangular numbers (see Table 9). From triangular numbers, other solid numbers arise – pyramidal numbers (*pyramis*) have their origin in triangular numbers, and from pyramidal numbers, we may derive other solid numbers, etc.[79]

---

79 Boethius, *Arith.* II, 21–23, pp. 129–135; cf. Nicomachus, *Arith.* II, 13, 2–14, 5, pp. 99–104.

Table 9 – Forming the sequence of plane numbers from triangular numbers

Triangular numbers (a_n)	(1)	3	6	10	15	21	28	etc.
$b_{n+1} = a_n + a_{n+1}$		1 + 3	3 + 6	6 + 10	10 + 15	15 + 21	21 + 28	etc.
Quadrangular numbers (b_n)	(1)	4	9	16	25	36	49	etc.
$c_{n+1} = a_n + b_{n+1}$		1 + 4	3 + 9	6 + 16	10 + 25	15 + 36	21 + 49	etc.
Pentagonal numbers (c_n)	(1)	5	12	22	35	51	70	etc.
$d_{n+1} = a_n + c_{n+1}$		1 + 5	3 + 12	6 + 22	10 + 35	15 + 51	21 + 70	etc.
Hexagonal numbers (d_n)	(1)	6	15	28	45	66	91	etc.
$e_{n+1} = a_n + d_{n+1}$		1 + 6	3 + 15	6 + 28	10 + 45	15 + 66	21 + 91	etc.
Heptagonal numbers (d_n)	(1)	7	18	34	55	81	112	etc.

This is how arithmetic understands the nature of numbers. The organising of numbers into geometrical shapes is not only in accordance with the real nature of numbers, but it also explains the organisation of reality since – in the words of Plato – numbers in their perfect geometrical form represent the four basic elements of everything created, i.e., the elements (fire is a regular tetrahedron, air is a regular octahedron, water is a regular icosahedron, and earth is a cube, i.e., a regular hexahedron) and the coherency of the entire cosmos (the divine layout of the universe, which is a regular dodecahedron).[80] It is evident that the basis for all these Platonic solids are triangles.

Geometry, in contrast, understands triangles (and geometrical shapes in general) differently. During Gerbert's times, the practical approach to geometry was dominant, as described, for example, by Martianus Capella in his allegorical compendium.[81] When Lady Geometry enters the wedding gathering of the celestials, she feels that she must explain why she is covered in dust (*pulvis*),

---

[80] Plato, *Tim.* 54d–55c. Up to the 12th century, the connection between Platonic solids and elements was almost unknown in Latin Christian Europe. The main reason is that Calcidius's translation of a commentary to Plato's *Timaeus* ends just before the theory is introduced in Plato's dialogue (i.e., since *Tim.* 53c). Nevertheless, there is little information about these solids and their identification with the elements – see, e.g., the somewhat unclear note in Isidore, *Etym.* III, 11, 2, or the reference to a pyramid (*pyramis*) as a representation of fire (*ignis*) in ibid. III, 12, 6, l. 4–6; cf. also Calcidius, *In Tim.* II, 326.

[81] Cf., e.g., Shelby, L. R., "Geometry." In: Wagner (ed.), *The Seven Liberal Arts...*, pp. 199–202.

for fear that she may be viewed as an unseemly, slatternly tramp (*indecenter squalentior peragratrix*). Her name refers to the fact that she has walked through (*permeare*) and measured (*admetiri*) the whole land (*tellus*), so she knows its shape (*forma*), magnitude (*magnitudo*), location (*locus*), and its various parts (*partes*) and sizes (*stadia*).[82] Thus, geometry is, first and foremost, geography, but it also describes celestial phenomena and the orbits of the heavenly bodies;[83] hence it is also cosmography. Therefore, it requires a knowledge of geometrical shapes and their properties, often related to practical surveying. Both Cassiodore and Isidore of Sevilla connect the emergence of geometry as a science to the flooding (*inundans*) of the river Nile in Egypt, when it was necessary to divide the land (*terra*) by various lines and measurements (*lineae et mensurae*).[84] This interpretation was, apparently, still popular in the times of Gerbert, since even *Geometria* (so-called *Geometria Gerberti*[85]), which is ascribed to him, presents a very similar explanation of the geometrical arts.[86] The very description of geometry stems from this – it is a science (*disciplina, scientia*) about measuring (*dimensiones*), which explores sizes (*magnitudines*) and shapes (*formae*). It is clear that its utility (*utilitas*) is priceless (*maxima*), since it helps reveal (*vestigare*) the beauty (*mirus*) of the Creation according to the plan of the Creator (*creator*), who arranged everything in accordance with the measurement (*mensura*).[87]

Not only through its references to the biblical *Book of Wisdom* but also through its building blocks (*elementa*), *Geometria Gerberti* complements the arithmetical art. However, although geometry explores the point (*punctum*), the line (*linea*), the surface (*superficies*), and solid objects (*soliditas*),[88] through which it follows up on the issue of figural numbers, unlike arithmetic, it does not inquire into the punctual representation of numbers (i.e., a quantity *per se*), but, in accordance

---

82 Martianus, *De nupt.* VI, 588, p. 206.
83 Ibid. VI, 580, p. 204.
84 Cassiodore, *Inst.* II, 6, 1, p. 150–151; and Isidore, *Etym.* III, 10, 1.
85 Cf. Materni, M., "Attività scientifiche di Gerberto d'Aurilliac." *Archivum Bobiense* 29 (2007), pp. 282–292; eadem, "La *Geometria Gerberti*: un manuale scolastico del X secolo." *Euphrosyne* 37 (2009), pp. 363–374; Guillaumin, J.-Y., "Les deux définitions de l'angle plan par Gerbert." In: Callebat, L. – Desbordes, O. (eds.), *Science antique – science médiévale (autour d'Avranches 235)*. Hildesheim: Olms-Weidmann, 2000, pp. 359–372; Beaujouan, G., "Les Apocryphes mathématiques de Gerbert." In: Tosi (ed.), *Gerberto – scienza, storia e mito...*, pp. 645–658; etc.
86 Gerbert [?], *Geom.* I, 1–2, pp. 48–50.
87 Ibid. I, 2–3, pp. 50–51; cf. *Sap.* 11, 20.
88 Gerbert [?], *Geom.* II, 1, p. 51.

with its practical application, it deals with the properties of geometrical objects (i.e., magnitude *per se*).

Gerbert wrote a letter concerning this very comparison of arithmetical and geometrical approaches to quantitative properties, which he addressed to Adelbold of Utrecht.[89] The letter was written during the last quarter of the 990s, probably when Gerbert held the office of archbishop in Ravenna,[90] possibly while preparing for departure to Rome, where Gerbert was to become Pope Sylvester II.[91] From the introduction of the preserved letter from Gerbert, it is clear that Adelbold was in contact with him – they had exchanged certain information about geometry.[92] From the context, it is apparent that Adelbold was astonished by the fact that the arithmetical and geometrical rules (*regulae*) for determining the area (*area*) of equilateral triangles (*trigoni aequilateri*) produced different numerical values for the same triangles.[93]

Firstly, Gerbert mentions that Adelbold knows the case of an equilateral triangle which has sides 30 feet long, with a height of 26 feet.[94] Geometry determines

---

89 Adelbold initially stayed in Liège (he was a disciple of Notker), then in Lobbes (where he co-worked with Heriger), and in the year 1010 he was promoted (through the influence of the emperor Henry II) to the position of bishop of Utrecht. He engaged in commenting of Boethius, and is the author of a text about the celestial sphere and texts about music, see, e.g., Wessely, O., "Adelbold von Utrecht und seine 'Musica'. *Anzeiger der Kaiserlichen Akademie der Wissenschaften. Philosophisch-historische Klasse* 24 (1949), pp. 575–583; Schmid, H., "Zur sogenannten 'Musica Adelboldi Traiectensis'. *Acta Musicologica* 28/2 (1956), pp. 69–73.

90 This letter is usually dated between the years 997–999 – see Gerbert, *Ad Adel.*, p. 41, which is also accepted by, e.g., the French translation of Gerbert's correspondence – cf. Gerbert, *Correspondence*, p. 701.

91 H. Pratt Lattin attempted to determine the time of writing more accurately, and she arrived at Gerbert's planned departure from Ravenna to Rome, where he became a pope shortly afterwards, i.e., February or March 999 – see *The Letters of Gerbert*, p. 301. Cf. also Rossi, P., "Sinossi delle principali differenti proposte di datazione." In: Gerbert, *Lettere (983–997)*. Transl. P. Rossi. Pisa: Plus – Pisa University Press, 2009, p. 205, although the dating for Gerbert's letter, acbribed to H. Pratt Lattin, contains a mistake or typo.

92 Gerbert, *Ad Adel.* 1, p. 43.

93 Ibid., pp. 43–44.

94 A triangle defined as such (*isopleurus*) can be found in, e.g., the so-called pseudo-Boethius's *Geometry II* where several other methods to determine its area are listed – see *Geom. II*, IV, 2–6, p. 148. Cf. the so-called *Geometria incerti auctoris*, where a different method of establishing the height of such a triangle is mentioned – *Geom. inc.* IV, 10, pp. 342–343.

the area of the equilateral triangle as half of the product of its side (*latus*) and height (*cathetus*),[95] which equals 390 square feet. This result is calculated correctly, but, as Gerbert points out, it uses inaccurate values. He refers to his earlier text[96] and states that the height of an equilateral triangle is more suitably defined as the length of one side of such a triangle, from which we subtract one-seventh of the length of the side.[97]

Therefore, according to Gerbert, we may learn the height of an equilateral triangle through the equation $v = 6/7a$, whereby v is the height of the triangle, and a is the length of the side of the same triangle. This rule is not accurate since the easiest way to determine the height of a given triangle, according to mathematical rules, is to use the equation $v = (\sqrt{3} : 2) \cdot a$, or via the Pythagorean theorem as $v^2 = a^2 - (1/2a)^2$. While we can say that according to the latter equations, the height of the triangle corresponds approximately to 0.866 times the side of the triangle (i.e., $v \approx 0.866a$), the ratio proposed by Gerbert only approximates this value ($v \approx 0.857a$).[98]

Consequently, Gerbert correctly claims that an equilateral triangle with sides 30 feet long does not have a height of 26 feet, as claimed by contemporary texts, but only approximates this value. Today, we would determine its height at approximately 25.714 feet, i.e., 25 and 5/7 feet. This leads to a different value assigned to the area of the given equilateral triangle. Contemporary geometric treatises suggest about 390 square feet, while Gerbert is justified in stating that the actual value is lower. However, the value determined by his rule is 385 and 5/7 square feet which is, paradoxically, more inaccurate than the original calculation of 390 square feet. Today, we can determine the area of this triangle to be approximately 389.711 square feet.[99]

---

95 An area of an equilateral triangle can be produced according to the equation S = (a · v) / 2.
96 It is not fully clear what text Gerbert refers to here – it could have been unpreserved texts about this topic or a different treatise dealing with the given topic. It is possible that the reference points towards *Geometry*, which is ascribed to him, and where we can find a detailed passage describing various types of triangles and ways we may determine their areas using the lengths of their sides, heights, etc. – see Gerbert [?], *Geom.* V, 1–VII, 1, pp. 71–97.
97 Gerbert, *Ad Adel.* 1, p. 44.
98 Cf. Miller, G. A., "Gerbert's Letter to Adelbold." *School Science and Mathematics* 21 (1921), p. 652.
99 Gerbert, *Ad Adel.* 1, p. 44.

Regardless of our possible claims that, according to geometry, the area of an equilateral triangle with a side 30 feet long is more than 385 square feet, less than 390 square feet, or precisely 390 square feet, the fact is that the arithmetical rule for determining the area of this triangle is entirely different. Gerbert states that, according to arithmetic, we do not need to know the height of a triangle – it is enough to know half the sum of the side length and its square root (i.e., $S_{Ar} = (a^2 + a) / 2$) and in this way, we arrive at a result of 465 feet,[100] which is the method described in contemporary texts concerning arithmetic and figural numbers.[101]

The reason for determining this value with the help of arithmetic is the fact that, according to figural numbers, an equilateral triangle with a side 30 feet long essentially represents the thirtieth triangular number (provided we consider the number 'one' as the first, since it is a potential triangle). As already stated, we determine the sequence of triangular numbers by adding the following natural number to the preceding triangular number, which means that the thirtieth triangular number is the sum of the values 1–30, equalling 465.[102]

Gerbert introduces another rule for determining a triangular number, related to its geometrical presentation (not unusual for figural numbers). A triangle with a side 30 feet long is expanded to a quadrangle, meaning that an identical triangle is added; its area is doubled and determined by multiplying one side by itself (i.e., the square root). In the case of these two triangles fitted into a quadrangle, one side overlaps (see Fig. 4); therefore, one side is added, and we arrive at the value of points for the given quadrangle. Finally, we turn the quadrangle back into a triangle by dividing the value by two.

---

100 Ibid. The equation is correct even in this case: $(30^2 + 30) / 2 = 930 / 2 = 465$.
101 Cf., e.g., *Geom. II*, V, 1–3, p. 149 – the example used here is a triangle with a side 28 feet long and its area is, according to arithmetic, 406 feet, for $(28^2 + 28) / 2 = 812 / 2 = 406$.
102 In essence: $\sum_{i=1}^{30} x_i$.

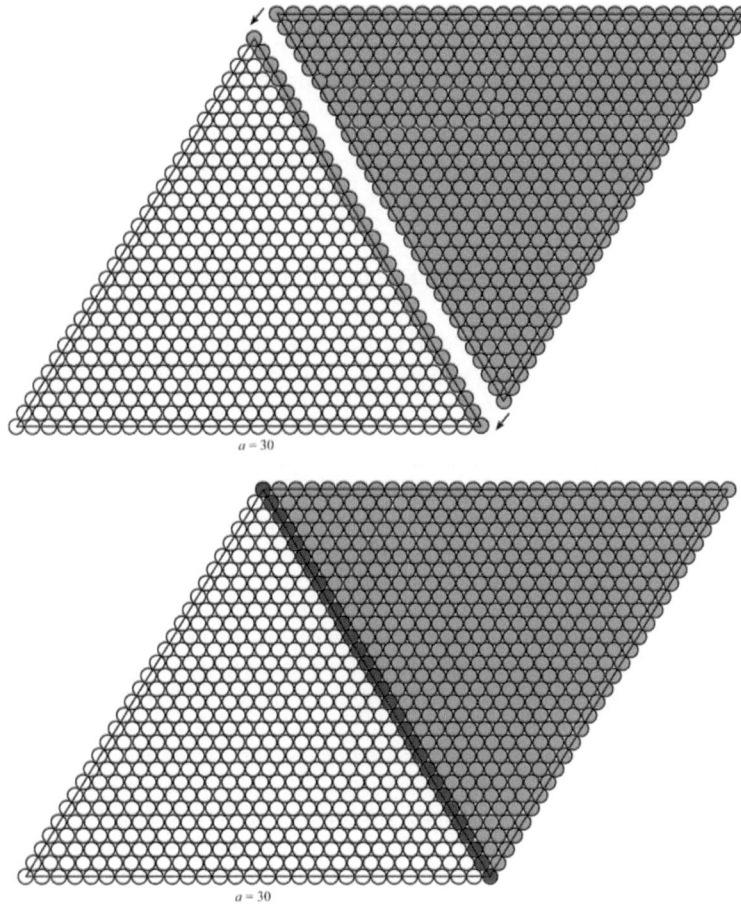

Fig. 4 – Point visualisation of the arithmetical equation for the area of an equilateral triangle, showing the necessity of adding the length of a side to its square root

In this way, Gerbert clearly demonstrates that the approaches of geometry and arithmetic to determining the area of a triangle are different and he, pursuing didactic goals, reiterates it in a more straightforward example – once again, it is an equilateral triangle; however, the length of its side is now seven feet. Geometry determines the area this way:

a) we determine the height[103] of the triangle – according to Gerbert, six-sevenths of a side length: $v = 6/7a$; and since $a = 7$, then $v = 6$;[104]
b) via the height, we arrive at the area if we multiply the length of the triangle by its height, and then divide the result by two: $S = (a \cdot v) / 2 = 21$.[105]

In contrast, arithmetic searches for the seventh triangular number (i.e., $\sum_{i=1}^{7} x_i$); therefore, the area of the given triangle is – with Gerbert's method – determined this way:

a) we determine the area of a quadrangle with the same side length: $a \cdot a$; and since $a = 7$, we get 49;
b) we add the length of sides, $a^2 + a$; thus, we get 56;
c) we divide the result by two, which equals 28.[106]

---

103 Here, Gerbert uses the less common term *perpendiculum*, which means perpendicular line; however, in this case, it is used with the same meaning as *cathetus*, i.e., the term for height. Explicit identification of both terms can be found in contemporary texts – see, e.g., a part of the so-called *Corpus agrimensorum*, which is ascribed to Aprofiditus and the architect Vitruvius Rufus (*Betrubus Rufus architecton*) – Epaphroditus–Vitruvius [?], *Excerpta* 4 (II, 1), p. 522. Cf. edition in Cantor, M., *Die römischen Agrimensoren und ihre Stellung in der Geschichte der Feldmesskunst. Eine historisch-mathematische Untersuchung.* Leipzig: Teubner, 1875, p. 208. For more about this text, see, e.g., Folkerts, M., "Mathematische Probleme im Corpus agrimensorum." In: Behrends, O. – Campogrossi Colognesi, L. (eds.), *Die römische Feldmeßkunst. Interdisziplinäre Beiträge zu ihrer Bedeutung für die Zivilisationsgeschichte Roms.* Göttingen: Vandenhoeck Ruprecht, 1992, pp. 319–321. For more about its familiarity and usage in Gerbert's time, see, e.g., Bergmann, W., "Gerbert von Aurillac und die Landvermessung." In: Junius, H. (ed.), *Ingenieurvermessung von der Antike bis zur Neuzeit.* Stuttgart: Wittwer, 1987, pp. 108–140; Shelby, L. R., "The geometrical knowledge of mediaeval master masons." In: Courtenay, L. T. (ed.), *The Engineering of Medieval Cathedrals.* London – New York: Routledge, 1997, pp. 32–33; or Folkerts, M., "Die Mathematik der Agrimensoren – Quellen und Nachwirkung." In: Möller, C. – Knobloch, E. (eds.), *In den Gefilden der römischen Feldmesser. Juristische, wissenschaftsgeschichtliche, historische und sprachliche Aspekte.* Berlin – Boston: De Gruyter, 2013, pp. 144–146.
104 The correct value (should we use listed equations with powers and roots) is slightly higher – $v \approx 6{,}062$ feet.
105 Should we calculate it with $v \approx 6.062$, then we arrive at the result $S \approx 21.218$ (square) feet.
106 The seventh triangular number (provided we include 'one' as the first potential triangular number) is actually 28 (see Table 8 and Fig. 3), which corresponds to the equation $\sum_{i=1}^{7} x_i = 28$ It can be expressed also as $1/2a\,(a+1)$ – cf. Miller, G. A., "The Formula $1/2a\,(a+1)$ for the Area of an Equilateral Triangle." *American Mathematical Monthly* 28 (1921), pp. 256–258.

Thus, it seems that an equilateral triangle with a side of the same length has two different values for its area, which is, of course, impossible (*nequit*);[107] therefore, Gerbert provides an explanation for the diverging values at the end of his letter. First, he reminds Adelbold that there are feet (*pedes*) of length (*longi*), of square (*quadrati*), and of cube (*crassi*).[108]

In *Geometria*, which is ascribed to Gerbert, this difference is also addressed using different terminology, and the inclusion of additional illustrative drawings. The first type is a linear foot (*pes linearis*), which expresses a measurement of length (*mensura linearis*), since it corresponds to one dimension, i.e., length (*longitudo*).[109] Similarly, in Boethius's *Introduction to Arithmetic*, the first example of a figural number is one that expands in only one direction, i.e., it possesses length exclusively, which means a linear number (*numerus linearis*).[110] According to *Geometria*, the second type is a superficial foot (*pes constratus*), which captures surface measurement (*mensura constrata vel plena*), whilst it tracks two dimensions – length and width (*longitudo et latitudo*).[111] Likewise, a plane number in arithmetic is a number that captures the movement of a line creating surface shapes.[112] A comparison of geometrical and arithmetical interpretations of length and square dimensions according to the listed texts is presented by Fig. 5. Finally, there is a cubic foot (*pes solidus*), which corresponds to a solid measurement (*mensura solida*), which captures also, in addition to length and width, height (*altitudo*) or thickness (*crassitudo*).[113] Therefore, a solid measurement includes three dimensions, which, according to arithmetic, corresponds to the movement of an area, as represented by solid numbers (*numerus solidus*).[114]

---

107 Gerbert, *Ad Adel.* 2, p. 45. As noted by M. Folkerts, to accurately establish the geometrical area of an equilateral triangle, we need to know its height, which we can determine via Pythagorean theorem; however, in this way we often get an irrational number, which could be problematic for practical surveying. Therefore, a triangular number would have been preferred, since it can always be expressed by a rational number, although it does not correspond to the geometrical area of a triangle, but describes something different – cf. Folkerts, "Die Mathematik der Agrimensoren…," pp. 138–140.
108 Gerbert, *Ad Adel.* 3, p. 45.
109 Gerbert [?], *Geom.* II, 5–6, pp. 55–56.
110 Boethius, *Arith.* II, 5, pp. 110–111.
111 Gerbert [?], *Geom.* II, 5–6, pp. 55–56.
112 Boethius, *Arith.* II, 4, pp. 108–109.
113 Gerbert [?], *Geom.* II, 5–6, pp. 55–56.
114 Boethius, *Arith.* II, 20, p. 129.

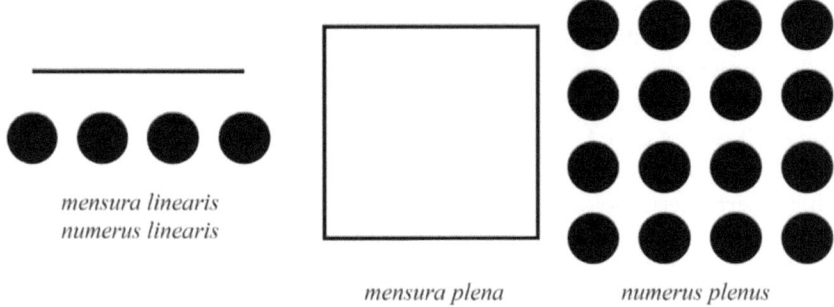

Fig. 5 – Length and superficial measurement ***versus*** linear and plane number

It is clear (as Gerbert briefly states) that when considering triangles, we must take plane numbers and square measures (*quadratus*)[115] into account. The arithmetical approach to triangles holds that every foot expresses a unit, i.e., one point. A sum of these units represents the value of a given number. Therefore, regarding arithmetic, every unit included in a specific number is taken as a whole (*integra*). In the case of the triangular number 28 (building upon Gerbert's drawing, which intentionally uses point representation via squares to highlight plane continuity) we can present this number as follows – see Fig. 6:[116]

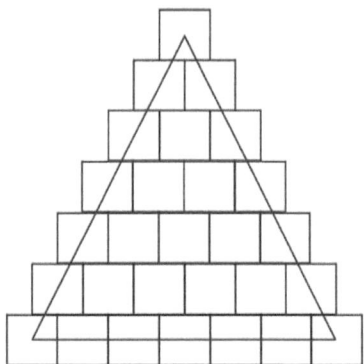

Fig. 6 – Triangular number 28 according to arithmetic, i.e., the arithmetical visualisation of an equilateral triangle with a side 7 feet long

---

115 Gerbert, *Ad Adel.* 3, p. 45.
116 Ibid.

Since arithmetic operates with every unit as a whole, regardless of whether the said unit (as seen by geometry) is wholly or only partly inside the given object, arithmetical understanding of geometrical shapes does not maintain a surface delineated by sides of a given shape; since the basis of all quantity is a unit. Thus, it is necessary to include every unit as a whole into a figurative expression of a given number.

However, geometry thinks very differently about a surface, since it is primarily interested in the area delineated by the sides (*latera*) of a specific shape. Gerbert reminds us that an equilateral triangle with a side 7 feet long must have a height of 6 feet and that allows us to create, through multiplication, a quadrangle, which is called "quadrangle-like" (*quasi quadratum*) by Gerbert, and which we can divide (*dimidare*) in order to determine the area of the triangle, i.e., 21 feet.[117] For a more precise visualisation, see Fig. 7. At first glance, it is a different shape from the one represented by an arithmetical figural presentation.

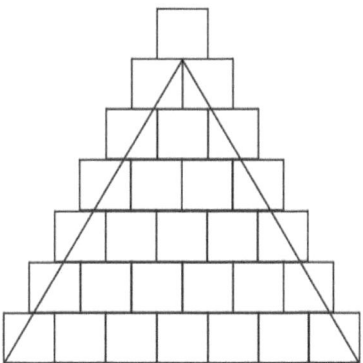

Fig. 7 – Equilateral triangle with a side 7 feet long according to geometry

Interestingly, Gerbert proposes another method to calculate the triangle's geometrical area based on this illustration. Since geometry focuses on a space delineated by the sides of an object, it is necessary to count as whole units only those points (squares or square feet) that are entirely inside the triangle. In Fig. 8, these points are shaded in the darkest colour and there are 15 of them. To them, we must add those points (squares) which intersect the sides of the triangle and are

---

117 Ibid.

only partially included in the shape (in Fig. 8, shaded in a lighter grey colour). However, we do not add the whole squares but only their halves. The triangle's sides intersect 12 squares; therefore, we add six of them to the calculation of geometrical content. The sum is 21 square feet, which is the area of the given tri-

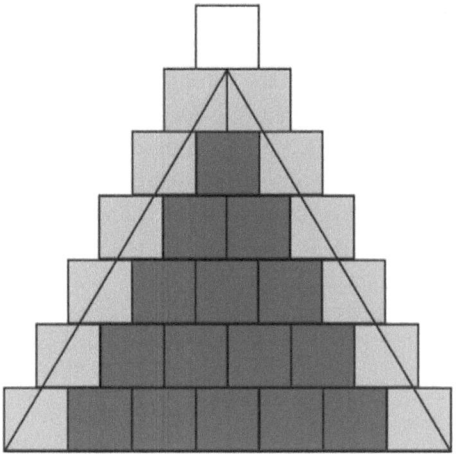

Fig. 8 – Calculating the area of an equilateral triangle according to geometry via a sum of square feet delineated by sides

angle according to geometrical rules.[118] Gerbert does not pause to mention that according to geometry, a square foot (point) not included in the shape delineated by sides (in Fig. 8, white square on the top) is not added to its area.

This concludes Gerbert's exposition of the different values pertaining to the area of the same equilateral triangle according to arithmetic and geometry. It is not only because both mathematical arts use different rules, but also because each of them comprehends the given triangle as a different object. This is reflected in the different focus of both arts. Geometry studies the material side of reality; therefore, it is interested in shapes inasmuch they are present in matter and how they delineate and structure matter. Geometrical shapes always concern exactly defined (sides, angles, etc.) parts of a specific segment in the material world, whose quantitative definitions it tries to capture. In contrast, arithmetic is concerned with form and quantitative value, leading to the uncovering of the

---

118 Ibid.

formative principle which is reflected in reality. It is closer to God's ideas and thoughts through which it can produce general quantitative findings that are realised in the material world.

Should we take these facts into account, then we can easily understand that in arithmetic the area of an equilateral triangle is the sum of units that form a given figural number, generally represented as a visualised form of a given quantity. Geometry, on the other hand, strives to precisely express the area of a shape, as defined by its sides.

3. Music and Harmony (Two Letters to Constantine)

When summarizing Pythagorean teachings, Aristotle stresses the importance of numbers and harmony for their ability to explicate a visible and a non-visible reality: numbers are the beginning of all things and represent the fundamental basis of individual entities. They determine time and justice, delimit the course of this world, and determine the principles of the cosmos. Numbers are mutually interconnected in ratios, creating a harmony that pervades the entire universe. The harmony resonates through the music of the spheres that permeates and echoes in the entire cosmos, and together with the number, it represents a fundamental metaphysical structure of reality as a whole.[119] Therefore, since pre-Socratic Antiquity, music has been closely linked to the art of mathematics; its importance does not lie only in its euphony and aesthetic-ethical effects – it represents the metaphysical order of reality that has the power to divert human beings from their sensual and changeable world and raise them closer to the divine origin.[120]

---

119 Aristotle, *Met.* I, 5, 985b.
120 Cf., for example, Rowett, C., "Philosophy's Numerical Turn: Why the Pythagoreans' Interest in Numbers is Truly Awesome." In: Sider, D. – Obbink, D. (eds.), *Doctrine and Doxography. Studies on Heraclitus and Pythagoras*. Berlin – Boston: De Gruyter, 2013, pp. 3–31; Šíma, A., *Svět vymezený a neomezený. Principy přírody ve filosofii Filoláa z Krotónu a u raných pythagorejců*. Červený Kostelec: Pavel Mervart, 2012, pp. 191–207; Burkert, W., *Weisheit und Wissenschaft. Studien zu Pythagoras, Philolaos und Platon*. Nürnberg: Hans Carl, 1962; Cherniss, H., *Aristotle's Criticism of Presocratic Philosophy*. New York: Octagon Books, 1971; Huffman, C. A., "Two Problems in Pythagoreanism." In: Curd, P. – Graham, D. W. (eds.), *The Oxford Handbook of Presocratic Philosophy*. Oxford: Oxford University Press, 2008, pp. 284–304; Cornelli, G., *In Search of Pythagoreanism. Pythagoreanism as an Historiographical Category*. Berlin – Boston: De Gruyter, 2013, pp. 137–187; etc.

Gerbert's outstanding knowledge of the musical art was a significant catalyst for his individual career, as is apparent in Richer of Reims' account of Gerbert's visit to Rome in 970. As a member of the Catalan delegation, Gerbert met Pope John XIII and, later, also Emperor Otto I. Both were attracted to his knowledge of music and astronomy, as neither of these sciences was deeply developed in Christian Europe (including Italy) at this time.[121] Music, therefore, represents Gerbert's stepping stone on his way to the emperor's court, and his cooperation with the Ottonian dynasty in many ways shaped his further career.

Therefore, when considering the importance of number theory to Gerbert, we cannot overlook the relation between music and musical theory to arithmetic (and metaphysics). While arithmetic, as stated before, inquiries into *multitudo per se*, i.e., a number; music is concerned with the relative properties of a number, i.e., numerical ratios. Gerbert described some aspects of his music teaching during the end of the 970s in two short texts – letters written to Constantine of Fleury and Micy, who is praised as a great expert on music in Gerbert's later letters.[122]

Both of Gerbert's musical letters to Constantine are brief commentaries on selected passages of Boethius's *Introduction to Music*. Their author, in fact, does not deal with the overlapping musical, philosophical, or metaphysical contexts of the topic discussed. In both cases, he holds strictly to mathematical explanations and, in theory and with the aid of practical examples, he expounds arithmetical operations with numerical ratios (musical intervals), or classifies the results of performed calculations.

This approach corresponds with Gerbert's note written at the beginning of his treatise called *Rogatus a pluribus*. He dissociates himself from auditory (or generally empirical, based upon feelings aroused by music) principles of musical theory as presented by Aristoxenus of Tarentum[123] and, instead, unambiguously

---

121 Richer, *Hist.* III, 44, p. 192. Cf. Lindgren, *Gerbert von Aurillac...*, pp. 69–72; or Flusche, *The Life and Legend...*, pp. 18–20.
122 Gerbert, *Ep.* 92, pp. 121–122; cf. Dachowski, *First Among Abbots...*, p. 221.
123 For more details see, e.g., Barker, A., "Music and Perception: A Study in Aristoxenus." *The Journal of Hellenic Studies* 98 (1978), pp. 9–16; Gibson, S., *Aristoxenus of Tarentum and The Birth of Musicology*. New York – Abingdon: Routledge, 2005; Huffman, C. A. (ed.), *Aristoxenus of Tarentum. Discussion*. New Brunswick: Transaction Publishers, 2012; Lichfield, M., "Aristoxenus and Empiricism: A Reevaluation Based on His Theories." *Journal of Music Theory* 32/1 (1988), pp. 51–73; Gurd, S. A., *The Origins of Music Theory in the Age of Plato*. London – New York: Bloomsbury Academic, 2019, pp. 91–155; and so on.

follows the Pythagorean mathematical structure of the musical art, including the tuning of organ pipes (*mensura fistularum*).[124] Since Gerbert deals exclusively with mathematics related to music science in his musical letters, this chapter will stress not only the mathematical context of his explanations but will also focus on the wider context of the understanding of music as a science and an art in the Early Middle Ages, and the philosophical consequences of this.[125]

We can assume that both of Gerbert's commentaries on Boethius's *Introduction to Music* were written as answers to questions raised by Constantine. Gerbert's responses in both cases are of a very similar structure: firstly, he turns his attention to a fragment of Boethius's text dealing with the relevant problem, followed by a quotation from *Introduction to Music* accompanied by a brief explanation. Gerbert presents specific mathematical examples to clarify the quoted passage (or Boethius is referred to again to demonstrate his identical understanding of the specific problem), whereupon the entire answer is summarized in general.

In the case of the commentary on Boethius's *Introduction to Music* II, 10 and IV, 6, the question concerns the doubling of superparticular ratios, and the classification of the products. Boethius states that, in fact, the product of the superparticular ratio and number 'two' does not result in a multiple or superparticular ratio.[126]

As an example Gerbert uses the sesquialter (i.e., a perfect fifth, or *diapente*, ratio 3 : 2). When the interval defined by the ratio 3 : 2 is doubled (*binario multiplicetur*), it is obvious that a three numerical sequence will be created (e.g., numbers 4 – 6 – 9). Between the first and the second member of this sequence (numbers 4 – 6) the superparticular ratio (*superparticularis proportio*) is 3 : 2 (a perfect fifth), equally, between the second and the third member of the sequence (i.e., numbers 6 – 9) is the same superparticular ratio (i.e., a second perfect fifth). Two perfect fifths (i.e., a doubled fifth) have, for example, the form of the numerical sequence 4 – 6 – 9. However, between the first and the third number of this numerical sequence (i.e., numbers 4 and 9), there is not

---

124 Gerbert, *Rogatus* 1–8, pp. 59–60; cf. Sigismondi, C., "Gerberto e la misura delle canne d'organo." *Archivum Bobiense* 29 (2007), pp. 355–396; or Williams, P., *The Organ in Western Culture, 750–1250*. Cambridge: Cambridge University Press, 1993, pp. 276–285.
125 Cf., e.g., Pizzani, U., "The Fortune of the *De Institutione Muscia* from Boethius to Gerbert D'Aurillac: A Tentative Contribution." In: Masi (ed.), *Boethius and the Liberal Arts*, pp. 97–138.
126 Boethius, *Mus.* II, 10, p. 240: ...*si superparticularis proportio binario multiplicetur id, quod fit, neque superparticulare esse, neque multiplex.*

a superparticular or multiplex ratio – but a multiplex superparticular number (*multiplex superparticularis*) – the ratio 9 : 4 (*duplex sesquiquartus*).[127]

In the second musical letter to Constantine, Gerbert focuses on a passage from the *Introduction to Music* II, 21 in which Boethius mentions two consequent superparticular ratios. When the lesser one (the one immediately following) is subtracted from the larger of the two superparticular ratios, the resulting difference of this subtraction will be less than one half of the subtracted ratio (the subtrahend), as the duplex product will be less than the subtrahend. Boethius offers a specific example of subtracting a perfect fourth, or *diatessaron* (ratio 4 : 3), from a perfect fifth, or *diapente* (ratio 3 : 2). The result of this calculation is a whole tone – in other words, a major second (*epogdous* or *tonus*, i.e. ratio 9 : 8). The duplex of the whole tone (ratio 81 : 64) is then a semitone less than a perfect fourth (ratio 4 : 3).[128]

Gerbert demonstrates these relations through an example of the smallest numerical sequence that expresses a ratio between a perfect fourth (*sesqualtera proportio*) and a perfect fifth (*sesquitertia proportio*): 6 – 8 – 9. From a perfect fifth (i.e., a *diapente* interval, defined by numbers 6 – 9) he subtracts a perfect fourth (i.e., a *diatessaron* interval, defined by numbers 6 – 8). The result is an interval between numbers 8 – 9 (*sesquioctava proportio*), i.e., a major whole tone (*tonus*). The double of this tone can be expressed, for example, by the three-member numerical sequence 64 – 72 – 81. While here (exactly as in the first of Gerbert's letters) the first and the second member of this sequence (like its second and third members) have the very same ratio in between (9 : 8, in this case). The ratio between the first and the third member of this sequence (i.e., between numbers 64 and 81) is less than a 4 : 3 ratio. The obvious conclusion is that the 9 : 8 ratio is less than half of the 4 : 3 ratio, which is – as Gerbert notes – valid for all results of subtraction of two consequent superparticular ratios.[129] In his letter, this statement is supported by two schemes demonstrating this rule on

---

127 Gerbert, *1 In Boeth. Mus.*, p. 29.
128 Boethius, *Mus.* II, 21, p. 254: *Ab omni superparticulari, si continuam ei superparticularem quis auferat proportionem, quae est scilicet minor, id, quod relinquitur, minus est ejus medietate, quae detracta est, proportionis: ut in sesqualtera vel sesquitertia. Quoniam sesqualtera major est, sesquitertiam de sesqualtera detrahamus. Relinquitur sesquioctava proportio, quae duplicata non efficit integram sesquitertiam proportionem, sed ea distantia minor est, quae in semitonio repperitur. Quod si duplicata sesquioctava comparatio non est integra sesquitertia, simplex sesquioctava non est sesquitertiae proportionis plena medietas.*
129 Gerbert, *2 In Boeth. Mus.*, p. 31.

the results of subtracting a perfect fourth from a perfect fifth and a major third (5 : 4) from a perfect fourth (4 : 3)[130] – see Fig. 9.

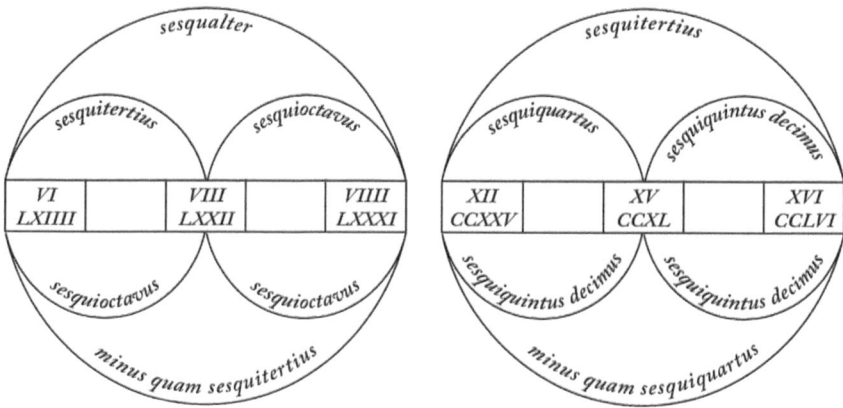

Fig. 9 – Subtraction of two consequent superparticular ratios according to Gerbert's schemes

As Gerbert follows Boethius's (and the Pythagorean) mathematical fundament of musical theory,[131] it is natural that his musical letters mainly have the character of a brief mathematical tractate presenting selected topics related to arithmetical operations with fractions (ratios). Although, at first sight, they seem to deal with sophisticated computations, they are elementary operations.

In the first of his letters, Gerbert considers doubling the ratios, and as an example, he presents the superparticular number 3 : 2. Mathematically we can express Gerbert's example in the following way:

$2 \cdot (3 : 2) = 2 \cdot (3 / 2) = 3^2 / 2^2 = 9 / 4 = 9 : 4.$

From this calculation we can easily deduct that the result of a perfect fifth doubling does not represent a multiplex ratio (i.e., when the lesser number is contained in the greater one more then once and fully corresponds with the value of the greater number; for example, the ratios 2 : 1; 3 : 1; 4 : 1; and so on;

---

130 Ibid., Tab. 1 (ad p. 31).
131 For a brief overview, see, e.g., Chadwick, H., *Boethius. The Consolations of Music, Logic, Theology, and Philosophy.* Oxford: Clarendon Press, 1981, pp. 88–90.

therefore double, triple, quadruple etc.) or a superparticular ratio (i.e., when the lesser number is contained in the greater number once but a certain part of the lesser number remains to make up the full value of the greater number; the remaining part can be expressed as a unit fraction of the lesser number, i.e., a fraction for which the number 'one' is the numerator; for example, the ratios 3 : 2; 4 : 3; 9 : 8; and so on, i.e., *diapente, diatessaron, tonus* etc.), but it is the multiplex superparticular (i.e., when the lesser number is contained in the greater one more than once, but a certain part of the lesser number remains to make up the full value of the greater number; the remaining part can be expressed as a unit fraction, i.e., fraction for which the number one is a numerator; for example, the ratios 5 : 2; 7 : 3; 9 : 4; and so on).[132]

If we wish to express the entire calculation as an interval visualized on a numeral axis, it can be demonstrated as follows (see Fig. 10):

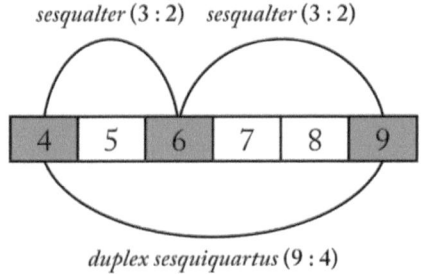

Fig. 10 – Duplex of the superparticular ratio 3 : 2 gives the result of the multiplex superparticular ratio 9 : 4; based on Gerbert's example

The result of the doubled perfect fifth is the 9 : 4 ratio; therefore, it is not a multiplex or a superparticular number, but it is a multiplex superparticular. It is necessary to note that a multiplex superparticular is not the result of doubling any superparticular numbers but only of a perfect fifth. When other superparticular ratios are doubled (for example, a perfect fourth, a whole tone, i.e., a major second, and so on), the result is always a superpartient ratio (i.e., the relationship between two numbers in which the lesser number is contained in the greater number, and in order to reach the value of the greater number, it is necessary to add a certain proportion of the lesser number that cannot be expressed by a unit

---

132 For more details see p. 39 above.

fraction of the lesser number).[133] Thus a doubled perfect fourth (4 : 3) represents a minor seventh (16 : 9) and a doubled major second represents a major third (81 : 64), and so on.

While Gerbert does not deal with the importance of the multiplication of a perfect fifth in his letter,[134] it is possible to assume that Constantine was aware of it. Therefore, Gerbert only reacts to the arithmetical part of the problem, since the musical one seems clear. What is, then, the importance of the double perfect fifth (alternatively its multiple or division) in music theory? It is a perfect fifth and its powers stand behind the Pythagorean fundamentals of mathematically expressible tuning based on the mutual relations between an octave (*diapason*, ratio 2 : 1) and a fifth (ratio 3 : 2).[135] When there are defined relative relations between the tones within an octave (expressed in a modern way, for example, from C_1 to C_2), the multiplication and division by a perfect fifth and an octave is used. The fifth ratio (*diapente*, 3 : 2) can be obtained by multiplication of a perfect prime (1 : 1, in the basic major scale, the *C* tone, i.e., 'Do') while we get the *G* tone (the so called the perfect fifth, i.e., 'Sol'):

$$(1:1) \cdot (3:2) = (1/1) \cdot (3/2) = (1 \cdot 3) / (1 \cdot 2) = 3/2 = 3:2.$$

The perfect fourth (*diatessaron*, ratio 4 : 3) can be mathematically obtained by dividing a perfect prime by a perfect fifth (the resulting tone would be one fifth less than the basic tone) and, consequently, when this proportion is multiplied by an octave (the ratio 2 : 1), the *F* tone (i.e., 'Fa') is obtained:

$$(1:1) / (3:2) = (1/1) / (3/2) = (1 \cdot 2) / (1 \cdot 3) = 2/3 = 2:3;$$
$$(2:3) \cdot (2:1) = (2/3) \cdot (2/1) = (2 \cdot 2) / (3 \cdot 1) = 4/3 = 4:3.$$

A major second (a major whole tone; i.e., *epogdous* or *tonus*) is defined as the doubled perfect fifth divided by an octave (*diapason*), and the *D* tone (i.e., 'Re') is obtained:

---

133 Cf. p. 39 above.
134 On Gerbert's teaching methods in the field of music see Richer, *Hist.* III, 49, p. 195. Furthermore, see, for example, Sigismondi, "Gerberto e la misura…," pp. 355–396; Nuvolone, F. G., "'Gerbertus musicus' e le attività culturali bobbiesi dell'annata 2002." *Archivum Bobiense* 24 (2002), pp. 7–48; Huglo, M., "Gerbert, théoricien de la musique, vu de l'an 2000." *Cahiers de civilisation médiévale* 43/170 (2000), pp. 143–160; Sachs, "Gerbertus cognomento musicus…," pp. 257–274; etc.
135 Cf., e.g., Rimple, "The Enduring Legacy…," pp. 455–457; or Bower, C. M., "The Modes of Boethius." *The Journal of Musicology* 3/3 (1984), pp. 252–263.

$2 \cdot (3:2) = 2 \cdot (3/2) = 3^2 / 2^2 = 9/4 = 9:4;$
$(9:4) / (2:1) = (9/4) / (2/1) = (9 \cdot 1) / (4 \cdot 2) = 9/8 = 9:8.$

In a similar way, a major third (the *E* tone, i.e., 'Mi'), a major sixth (*A* tone, i.e., 'La') and a major seventh (the *B* tone, i.e., 'Si') can be obtained.[136] An obvious reason why Constantine asked Gerbert to explain the way to double superparticular ratios is that the mutual multiplication and division of the ratios 2 : 1 and 3 : 2 allow us to define the scale of tones in full.

The second of Gerbert's musical letters is also concerned almost exclusively with arithmetical operations with ratios, i.e., music intervals. In this case, it deals with subtraction and multiplication by two. Gerbert's example shows that the result of subtraction of the two consequent superparticular ratios is less than one half of the subtracted ratio, as the doubled result of this subtraction is less than the entire subtracted ratio. Modern mathematical notation expresses it in this way:

$(3:2) - (4:3) = (3/2) - (4/3) = (3/2) \cdot (3/4) = (3 \cdot 3) / (2 \cdot 4) = 9/8 = 9:8;$
$2 \cdot (9:8) = 9^2 / 8^2 = 81/64 = 81:64;$
$(4/3) - (81:64) = (4/3) \cdot (64/81) = 256/243 = 256:243;$
$81:64 < 4:3.$

Expressed in musical terms: if we subtract a perfect fourth from a perfect fifth, a major second (major whole tone) is obtained and two major seconds (two whole tones) are one *limma* less than a perfect fourth. From the mathematical point of view, it basically means that if we subtract from the ratio 3 : 2 (for example, the interval determined by the numbers 6 – 9, or 192 – 288) the ratio 4 : 3 (for example, the interval determined by the numbers 6 – 8, or 192 – 256) it will result in the ratio 9 : 8 (for example, the interval determined by the numbers 8 – 9, or 256 – 288). If the 9 : 8 ratio is doubled, we obtain the 81 : 64 ratio (for example, the numeral sequence 64 – 72 – 81, or 192 – 216 – 243).[137] The 81 : 64 ratio, i.e., the 9 : 8 ratio doubled, is then less than the 4 : 3 ratio (for example, the interval determined by the numbers 63 – 84, or by the numbers 192 – 256) by the 256 : 243 ratio as more illustratively demonstrated in the following diagram (see Fig. 11).

---

136 For more details, see, e.g., Meyer, C., "*Gerbertus musicus*: Gerbert et les fondements du système acoustique." In: Charbonnel, N. – Iung, J. E. (eds.), *Gerbert l'européen. Actes du colloque d'Aurillac 4–7 juin 1996*. Aurillac: Société des lettres, sciences et arts La Haute Auvergne, 1997, pp. 183–192; or Nuvolone, F. G., "Gerberto e la musica." *Archivum Bobiense* 5 (2005), pp. 145–164.

137 Cf., for instance, Chadwick, *Boethius…*, pp. 88–89.

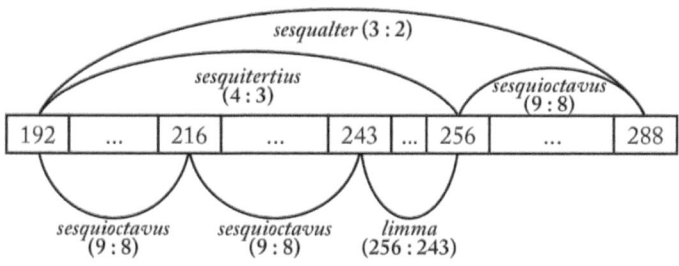

Fig. 11 – Duplex of the difference between a perfect fifth and a perfect fourth is less than a perfect fourth; based on Gerbert's example

Also, in this case, Gerbert does not stress the importance of the presented calculations to musical theory itself. The size of a whole major tone interval (a major second, i.e., the ratio 9 : 8) can be obtained through multiplication and division of a perfect fifth by an octave (as demonstrated with the help of Gerbert's first letter). Still, it can also be defined with the use of subtraction of a perfect fourth from a perfect fifth. This way, we can also determine the firm mutual relationship between a perfect fifth and a perfect fourth, or between a perfect fourth and a major second. Fig. 11 demonstrates that, while a perfect fifth represents the sum of a perfect fourth and a whole major tone, a perfect fourth is the sum of two whole tones and one semitone (*limma*).

That is to say, since an octave is the sum of a perfect fifth and a perfect fourth, within one octave the distances between the individual tones (C_1 to C_2) are determined by the intervals of five tones (ratios 9 : 8) and two semitones (between the tones *E* and *F* and between the tones *H* and C_2; in other cases there are intervals of whole tones). The mathematical formula of these musical relations can be presented, for example, in the following way:

(2 / 1) = (4 / 3) + (3 / 2) = (4 · 3) / (3 · 2) = 12 / 6 = 2 : 1;
(4 / 3) = (9 / 8) + (9 / 8) + (256 / 243) = (9 · 9 · 256) / (8 · 8 · 243) = 20 736 / 15 552 = 4 : 3;
(3 / 2) = (4 / 3) + (9 / 8) = (4 · 9) / (3 · 8) = 36 / 24 = 3 : 2;
(2 / 1) = 5 · (9 / 8) + 2 · (256 / 243) = (9^5 / 8^5) + (256^2 / 243^2) = (59 049 / 32 768) + (65 536 / 59 049) = (59 049 · 65 536) / (32 768 · 59 049) = 3 869 835 264 / 1 934 917 632 = 2 : 1.

This fact also plays an important role in musical theory, since it defines the tones (sounds) of an octave with the help of whole tones and semitones.

If we decide to follow Gerbert's (and Boethius's) example and subtract another two consequent superparticular ratios, i.e., the ratio 5 : 4 from the ratio 4 : 3, the result will be an interval of 16 : 15 (i.e., a lesser diatonic semitone). Its doubled quantity will provide a ratio of two semitones or a so-called diminished third (ratio 256 : 225). Differences between other consequent superparticular ratios, thus, also provide various small intervals used in music theory.

With its theory based on auditory perception or on mathematical fundamentals, in the Middle Ages, music was considered part of the liberal arts. Gerbert does not discuss the reason for his interest in music or the importance of this art in his letters to Constantine; however, the traditional definition of music as understood in the Middle Ages from late antiquity and explicitly defined in, for example, Augustine's work *De musica*, is closely tied to melody making, i.e., knowledge of the proper measure.[138] We often come across this definition during late antiquity or the early Middle Ages, perhaps in minor variations.[139] In his dialogue *De ordine*, Augustine thoroughly explains the systematic taxonomy of the liberal arts. He places music in first place among the mathematical sciences. The prestigious status of music is well deserved, since music represents a certain form of transition between sensory perception and purely rational cognition. Auditory perception holds an irreplaceable position within the musical art, as it naturally informs us about sounds, tones, melodies, rhythm, etc. Sensory impulses can bring pleasure or displeasure; however, various consonances and harmonies do not originate in our hearing of them but in the relationships of numbers to other numbers (i.e., ratios). Therefore, numbers are always and necessarily the basis of music in any of its forms. As numbers derive from God, it is impossible to doubt the divine origin of music. These numbers prove the divine perfection and represent the harmony that created and organized the world.[140]

---

138 Augustine, *Mus.* I, 2, 2: *Musica est scientia bene modulandi.* Cf., e.g., Censorinus, *Die nat.* X, 3, p. 16. Cf. Mathiesen, T. J., *Apollo's Lyre. Greek Music and Music Theory in Antiquity and the Middle Ages.* Lincoln – London: University of Nebraska Press, 1999, pp. 614–616, 619–622.

139 See, for example Martianus, *De nupt.* IX, 930, p. 356; Cassiodore, *Inst.* II, 5, p. 143; Isidore, *Etym.* III, 15; etc.

140 Augustine, *Ord.* II, 14, 39–41. Cf., e.g., Bai, J., "The Spectrum of the Divine Order: Goodness, Beauty, and Harmony." *Soundings: An Interdisciplinary Journal* 102/1 (2019), pp. 1–30; Harrison, C., "Augustine and the art of music." In: Begbie, J. S. – Guthrie, S. R. (eds.), *Resonant witness. Conversations between Music and Theology.* Grand Rapids – Cambridge: William B. Eerdmans, 2001, pp. 27–45; La Croix, R. R. (ed.), *Augustine on Music: An Interdisciplinary Collection of Essays.* Lewiston: Edwin Mellen Press, 1988; Bai, J., "Numbers: Harmonic Ratios and Beauty in Augustinian

As distinct from Augustine, Martianus Capella, in the compendium *De nuptiis Philologiae et Mercurii*, positions Music as the very last of all the liberal arts, representing the culmination of all knowledge. The arrival and presentation of Harmony (Music) at the wedding ceremony of the celestials is in some respects unique. Harmony waits at the entrance longer than all the other arts. The gods argue for a long time over whether they would like to hear anything more (some would rather start the wedding feast) and, although the performance of Lady Harmony is finally allowed, it is done so on the condition that presentations by the other two sister Muses – Architecture and Medicine – be abandoned.[141]

The entrance of Harmony ends all arguments and disputes between the gods over whether to listen to another lecture on the liberal arts or not. The presence of this art itself (naturally accompanied by strong visual and auditory effects) calms and astonishes everyone.[142] The harmonic and positive atmosphere persists during Harmony's presentation of the musical art,[143] watched enviously by the other sisters. Harmony impresses everyone by her noble appearance, the delightful rustling sound of her robe, and the sounds of her musical instruments (including a strange round shield interlaced with strings). The gods are so enchanted by Harmony and her music that they fully recognize the divine character of her performance, even giving her a standing ovation.[144] Martianus Capella presents music as superior to the gods – they honour her, she calms them with her melodies, brings them peace, prudence, harmony, and perfection. Music therefore reaches the ultimate heights.[145]

---

musical Cosmology." *Cosmos and History: The Journal of Natural and Social Philosophy* 13/3 (2017), pp. 192–217; Harrison, C., *On Music, Sense, Affect and Voice*. London – New York: Bloomsbury Academic, 2019; etc.

141 Martianus, *De nupt.* IX, 888–898, pp. 337–342; cf. Petrovićová, K., *Martianus Capella. Nauky „na cestě" mezi antikou a středověkem*. Brno: Host, 2010, pp. 34–36.

142 Martianus, *De nupt.* IX, 905, pp. 334–345; cf. Mathiesen, *Apollo's Lyre*…, pp. 625–627.

143 Martianus, *De nupt.* IX, 996, p. 384. For more on the presentation of the musical art according to Martianus Capella, see, e.g., accompanying texts in edition by L. Cristanteho – see *Martiani Capellae De nuptiis Philologiae et Mercurii liber IX*. Ed. and transl. L. Cristante. Padova: Editrice Antenore, 1987; early medieval reception during the Carolignian era is comprehensively covered by Teeuwen, M., *Harmony and the Music of the Spheres. The Ars Musica in Ninth-Century Commentaries on Martianus Capella*. Leiden: Brill, 2002.

144 Martianus, *De nupt.* X, 910, p. 347.

145 Cf., e.g., Moretti, G., "Harmonia allegorica: Il melos multiforme che fanda l'armonia del mondo nel *De nuptiis Philologiae et Mercurii* di Marziano Capella." *PAN. Rivista di Filologia Latina* 2 (2013), pp. 131–158.

Similarly, Isidore of Sevilla in *Etymologies* states that no knowledge (*disciplina*) is perfect without music, as without music, nothing would exist.[146] This is how the Pythagorean idea of the importance of music was disseminated across Europe – the entire universe and cosmos were created with the help of tonal scales, and everything celestial moves accompanied by the sound of cosmic harmony.[147] The cosmological and metaphysical foundations of the universe are determined by music (i.e., by numeral ratios). These numbers allow us to recognize the truth about the cosmos, and, at the same time, they touch humans and human souls. Sensory perception is adjusted by the rational cognition of numeral relations in a human soul.[148]

Boethius agrees with Augustine that music approaches us through our sense of hearing; however, harmony can engage the rational part of the human soul. It is the harmony reflected by the entire range of human behaviours, and which fundamentally affects the ethical activities of human beings. Therefore, music forms the character of people; however, beyond our misleading and unstable sensory (auditory) cognition, it is always necessary to seek true, stable, and permanent proportional mathematical relationships, which define the real value of music.[149]

The emphasis on music's mathematical grounding, typical in the Middle Ages, as shown, for example, in the *Lucidarium* by Marchetto of Padua,[150] probably influenced Gerbert's commentaries on Boethius's *Introduction to Music*. Proportional mathematical foundations (the determining aspect of music) form the order of the universe, cosmos, and humankind, including the journey towards Good and the Truth (in Boethius's terms *musica mundana* and

---

146 Isidore, *Etym.* III, 17; cf. Henderson, J., *The Medieval World of Isidore of Seville. Truth from Words*. Cambridge: Cambridge University Press, 2007, pp. 55–57.
147 Isidore, *Etym.* III, 17: *Nam et ipse mundus quadam harmonia sonorum fertur esse conpositus et caelum ipsud sub harmoniae modulatione revolvi*. Cf., for example, Wallis, F., "Isidore of Seville and Science." In: Fear, A. – Wood, J. (eds.), *A Companion to Isidore of Seville*. Leiden: Brill, 2019, p. 204.
148 Cf., e.g., Ribémont, B., "Isidore de Séville et les mathématiques." In: Baillaud, B. – De Gramont, J. – Hüe, D. (eds.), *Discours et savoirs: encyclopédies médiévales*. Rennes: Presses universitaires de Rennes, 1998, pp. 33–39.
149 Boethius, *Mus.* I, 1, pp. 179–187; cf., e.g., Schrade, L., "Music in the Philosophy of Boethius." *The Musical Quarterly* 33/2 (1947), pp. 188–200.
150 Herlinger, J. W., *The Lucidarium of Marchetto of Padua. A Critical Edition, Translation, and Commentary*. Chicago: University of Chicago, 1985, pp. 82–84. Furthermore see Celhoffer, M., "Hudba, animus a proporce v kontextu Marchettova Lucidaria aneb „svůdné vábení Sirén"." *Musicologica Brunensia* 46/1–2 (2011), pp. 49–54.

musica humana[151]). Therefore, naturally, the arithmetical aspect of music must be recognised in order for us to regard music as a science and knowledge rather than "merely" as a practical performance of musical art.

Thus, music was regarded by Gerbert as a liberal art, i.e., a knowledge that should lead us to the Truth; and what is more, it does not involve only the music perceived by our sense of hearing but, for example, the music of the spheres, which is inaudible.[152] It forms the very structure of the universe. Music stood at its beginning and defines the rules of this world (by numeral ratios, i.e., musical intervals), as it forms human beings themselves, to whom it opens the way to perfection in cognition, and in all their actions. An obvious result of perceiving music in this way is that it allows us to concentrate primarily on the mathematical description of fundamental musical relations.

As we have seen, Gerbert's commentary on Boethius's *Introduction to Music* follows the approach it does probably in answer to questions raised by Constantine of Fleury. Gerbert's typical emphasis on exemplary pedagogical explanation, and, at the same time, on the practical applicability of a given intellectual problem (e.g., in terms of music, his practically or pedagogically oriented activities related to organ making[153]) led him to the effort of explaining musical (and also metaphysical and cosmological) problems with the help of the simplest examples and models, using his considerable mathematical erudition to the best of his ability.

4. Astronomy and Timekeeping (Letter *De sphaera*, Richer's *Historia*, and the horological letter to Adam)

While music inquiring into a quantity (number) related to other numerical values provides knowledge about the order of the universe, when astronomy expands upon the arithmetical properties of numbers, it focuses on magnitude; however, unlike geometry, it does not examine the immutable and stable properties of shapes, but it applies geometrical art to descriptions of moving space objects and tries to capture the orbits of celestial bodies. Thus, astronomy enquires into the

---

151 Boethius, *Mus.* I, 1, pp. 187–189; cf. Mathiesen, *Apollo's Lyre*..., pp. 630–636; or Moyer, "The *Quadrivium*...," pp. 492–494.
152 Boethius, *Mus.* I, 9, p. 196; cf., for instance, Leach, E. E., *Sung Birds. Music, Nature, and Poetry in the Later Middle Ages*. Ithaca – London: Cornell University Press, 2007, p. 68.
153 Cf., for instance, Williams, *The Organ*..., pp. 283–285, or Flusche, *The Life and Legend*..., pp. 159–161.

order (based on numbers) through which God created the whole cosmos, whilst astronomical knowledge represents a way to recognise God's wisdom.[154]

Astronomy is a field that Gerbert's intellectual renown is deeply tied to. Here, also, Gerbert builds upon the evidenced contemporary interest in this science; however, in his case, we may notice a slight shift towards the practical use of astronomy, with a stress on pedagogical and applied aspects of this discipline. This might have been inspired by his stay in the Pyrenean foothills, where he became acquainted with a modified conception of astronomy, and by an interest in the utilisation of theoretical findings and available tools in everyday practice. Moreover, we may ascribe the re-introduction of the astrolabe in the Latin West to Gerbert, as well as a minor interest in the *computus*, which was very popular at that time.

We are well informed about Gerbert's pedagogical activities in Reims, thanks to Richer of Reims. Therefore, we can reconstruct a relatively clear picture of content and extent of the knowledge Gerbert was passing on to his students during the 970s in the cathedral school in Reims. It appears that he put the main emphasis on the practical utilisation of individual findings both in the case of *trivium* and *quadrivium* education. Concerning the former, Richer lists authoritative sources for dialectic (comprised mostly of traditional texts of so-called *logica vetus*, i.e., Porphyry, Aristotle, Marius Victorinus, Cicero, and Boethius[155]) and rhetoric (ancient poets or dramatists Virgil, Statius, Terentius, Juvenal, Persius, Horace, and Lucan[156]), and goes on to mention Gerbert's effort to apply theoretical knowledge to practical examples;[157] his tool for teaching rhetoric, for instance, i.e., twenty-six sheets of parchment fastened together, might serve as a demonstration of his practical activities.[158] Gerbert himself summarises his efforts in a letter in which he explicitly states that he wishes to follow in the footsteps of Cicero, and combine usefulness with nobility.[159]

---

154 See, for example, Augustine, *Ord.* II, 15, 42; or idem, *Lib. arb.* II, 16, 42; cf., e.g., McCluskey, S. C., "Astronomy in the Service of Christianity." In: Ruggles, C. L. N. (ed.), *Handbook of Archaeoastronomy and Ethnoastronomy*. New York: Springer, 2015, pp. 165–179.
155 Richer, *Hist.* III, 46–47, pp. 193–194.
156 Ibid. III, 47, p. 194.
157 Ibid. III, 48, p. 195.
158 Gerbert, *Ep.* 92, p. 121; cf. Lake, J., "Gerbert of Aurillac and the Study of Rhetoric in Tenth-Century Rheims." *The Journal of Medieval Latin* 23 (2013), pp. 61–63.
159 Gerbert, *Ep.* 44, p. 72; cf. Darlington, "Gerbert, the Teacher," pp. 456–476.

Similarly, while teaching the *quadrivium*, Gerbert employs multiple practical aids to expand and popularise knowledge of various operations related to these disciplines. For music, he used a monochord,[160] while for geometry and (apparently) arithmetic, he utilised an abacus[161] (as will be discussed later). The most detailed description of Gerbert's illustrative aids and tools, through which he led his disciples to an understanding of often complex subject matter, is provided by Richer with regard to astronomy. Since these devices (the celestial globe, observational hemisphere, and two armillary spheres) are supposed to represent the movement of stars, planets, or the Sun in the sky, we may say that these aids acted as a complete introduction to astronomy. Naturally, the introduction was simple, pedagogically-oriented, provided a primarily qualitative description of the universe, and, in comparison to contemporary astronomy in the Muslim cultural environment, Gerbert's tools seem like the most elementary propaedeutic to this art.

The very fact that Gerbert, during the last third of the 10th century, constructed and used these pedagogical, explanatory, and observational tools is somewhat untypical for Latin teachings of astronomy. Gerbert's own approach to astronomy was probably inspired by similar practices to which he was introduced during his studies in Catalonia, where Arabic astronomical teachings influenced contemporary Christian centres of knowledge. We have only a minimum of direct sources, which could help us state with certainty the subject of Gerbert's studies on the Iberian Peninsula. Richer states only that Gerbert's Catalan studies were related to *mathésis*, i.e., mathematics, including the four disciplines of the *quadrivium* (arithmetic, geometry, music, and astronomy).[162] Another contemporary source, the chronicle of Ademar of Chabannes, mentions that Gerbert studied wisdom (*sofia*) in Cordoba[163] (although this information is likely to be fallacious).[164] Later, Gerbert himself wrote several letters to his friends in the Pyrenees, but the only astronomical text he mentions is an unknown treatise *De astrologia* (possibly a treatise on the astrolabe), which is supposed to have been translated by deacon Lupitus of Barcelona.[165]

---

160 Richer, *Hist.* III, 49, p. 195.
161 Ibid. III, 54, p. 198.
162 Ibid. III, 43, p. 192.
163 Ademar, *Chron.* III, 31, p. 154.
164 Cf. Flusche, *The Life and Legend…*, pp. 111–116.
165 Gerbert, *Ep.* 24, p. 47. For other letters, addressed to the Pyrenees, see ibid. 25, pp. 47–48; or 112, pp. 140–141.

During the 12th century, when the negative legend of Gerbert formed, connecting his intellectual and political endeavours with the forces of the Devil, we can find more detailed information about Gerbert's studies in Catalonia. A comprehensive summary is provided by William of Malmesbury, who dedicated part of the second book of his work *Gesta regum Anglorum* to Pope Sylvester II. The astrological knowledge of the Saracens was supposedly the main reason for Gerbert's journey to the Iberian Peninsula, and the young student achieved great results: his skill with the astrolabe surpassed even that of Claudius Ptolemy, his astronomical knowledge was greater than the knowledge of the philosopher Alchandreus,[166] and even Julius Firmicus Maternus could not better him in the knowledge of astrology. Gerbert mastered both beneficial (i.e., mathematics, the *quadrivium*) and maleficent (i.e., black magic, especially necromancy) knowledge.[167] We cannot consider William's description to be a valid historical account, since it clearly reflects the writer's opinion of astronomy (and astrology) formed in a completely different cultural and historical context, and in pursuit of different goals. It probably has very little in common with the reality of Gerbert's stay on the Iberian Peninsula.

On the other hand, it seems likely that Gerbert drew his knowledge of astronomy mainly from Latin written sources, about which we have a comparatively clearer idea, since he himself mentions them on various occasions. Besides Christian thinkers of the 8th and the 9th century,[168] we can presuppose that Gerbert's knowledge of the Latin popular-science Aratian tradition of astronomy came from the treatise *De astronomia*, which is ascribed to the Roman scholar Gaius Julius Hyginus,[169] and from the astronomical and astrological text

---

166 Latin collection of originally Arabic (partially Jewish) texts on astronomy and astrology from the second half of 10th century, derived in part possibly from the works of Arabic scholar Al-Kindi; for further reading see Juste, D., *Les Alchandreana primitifs. Étude sur les plus anciens traités astrologiques latins d'origine arabe (Xe siècle)*. Leiden: Brill, 2007.
167 William, *Gesta reg.* II, 167, 1–3, p. 280.
168 The widely read texts of the Venerable Bede which inquire into astronomy are worth mentioning, e.g., *De temporum ratione, De temporibus liber*, or *De natura rerum*. Furthermore, it is also worthwhile listing the 9th century scholars: John Scotus Eriugena (especially his commentary to Martianus' *Marriage – Annotationes in Marcianum* or appropriate passages of *Periphyseon*), or Remigius of Auxerre (*Commentum in Martianum Capellam*).
169 Cf., for example, Van de Vyver, "Les plus anciennes Traductions…," p. 676; or Dekker, E., *Illustrating the Phaenomena. Celestial Cartography in Antiquity and the Middle Ages*. Oxford: Oxford University Press, 2013, pp. 198–201.

Astronomica by the Roman poet Marcus Manilius.[170] Gerbert also knew the work of Pliny the Elder,[171] and he was acquainted with the texts of Cicero *Scipio's dream*[172] and with Macrobius' commentary to it.[173] In addition, Calcidius' commentary to *Timaeus* by Plato was not unknown to him.[174] He might have been working with Boethius's unpreserved treatise on astronomy,[175] and he creatively elaborated on theories originating from the encyclopaedic tradition of late antiquity and the Early Middle Ages. In addition to the aforementioned Martianus Capella, he was definitely familiar with the compendiums of secular teachings written by Cassiodore and Isidore of Sevilla.[176]

Even this brief list clearly indicates that Gerbert was well versed in contemporary available Latin texts on astronomy. However, during his studies in Catalonia, he became acquainted with astronomy as a scientific discipline whose findings were not derived exclusively from books but often also from empirical observation.[177] He tried to include this practical element in his lessons on astronomy.

This aforementioned "astronomical bibliography" (possibly combined with the manuscripts of Carolingian scholars) might represent the main theoretical background of astronomy as known to Gerbert and his contemporaries. Richer's description of the teacher's tools does not provide an analysis of Gerbert's interpretation of the cosmos, which these instruments were required to elucidate. However, based on the listed (mostly) Latin sources, we may with relative ease reconstruct the fundamental theses through which students in Reims were initiated into the mysteries concerning the constitution and structure of the universe, celestial movements, etc. In this way, we may reconstitute the specific framework for a Latin introduction to geocentric astronomy, as presented by Gerbert at the end of the 10th century.

---

170 Cf. Gerbert, *Ep.* 130, p. 158. It is not clear whether Gerbert actually means Marcus Manilius or Boethius – for various interpretation see Lindgren, *Gerbert von Aurillac...*, pp. 36–37; or Goold, G. P., "Introduction." In: Manilius, M., *Astronomica*. Ed. and trans. G. P. Goold. Cambridge: Harvard University Press, 1977, p. cviii.
171 Gerbert, *Ep.* 7, p. 30.
172 Ibid. 167, pp. 195–196.
173 Cf., e.g., Zuccato, "Gerbert of Aurillac...," p. 758.
174 Gerbert [?], *Geom.* II, 6, p. 56.
175 Gerbert, *Ep.* 8, p. 31.
176 Ibid. 168, p. 197.
177 Cf. Saliba, G., *Islamic Science and the Making of the European Renaissance*. Cambridge: MIT Press, 2007; or Zuccato, "Gerbert of Aurillac...," pp. 742–763.

The first item Richer introduces in his account is Gerbert's celestial sphere, i.e., a celestial globe made from wood and serving as a model of the world sphere (*mundi speram*).[178] Through this model, a student could become acquainted with the world sphere and the fundamental order of the cosmos.

The celestial sphere represented the outermost boundaries of the universe as a whole. During the Early Middle Ages, the universe (*mundus*), known as cosmos by the Greeks, was viewed as a universality consisting of the sky (*caelo*) and the Earth (*terra*).[179] The sky is round, it turns, and it has stars imprinted on itself.[180] Therefore, the celestial sphere is rounded and the Earth is located in its centre.[181] The celestial globe shows the world sphere as viewed from the outside, with the Earth located in the middle.[182] Our Earth is of negligible size compared to the sky;[183] thus, it can represent the point that is the centre of the universe.[184]

Gerbert's globe was further accompanied by a horizon (*orizon, limitans, determinans*) and the whole sky model was tilted following the axis of the world sphere that went through the two poles of the sphere (*duo poli*) – the northern and the southern.[185] The world sphere rotates (from east to west) during the period of one day.[186] This movement is the rotation of the sphere around its axis, which cuts the world sphere at the northern (*Boreus*) and the southern (*Austronotius*) pole.[187] The main points of the celestial globe are delineated by the centre, axis, and poles (see Fig. 12, in which the Earth is made more prominent for illustrative purposes).

---

178   Richer, *Hist.* III, 50, pp. 195–196.
179   Cf. Isidore, *Etym.* III, 29, 1; Bede, *De nat. rerum* 3, p. 194.
180   Cf. Isidore, *Etym.* III, 31, 1.
181   Cf. ibid. III, 32, 1; Abbo, *De spere*, p. 120; etc.
182   Cf. Martianus, *De nupt.* VIII, 814, p. 309.
183   The size of the Earth was well-known to medieval authors – they took over values estimated by Eratosthenes and they understood the methods of its measurement – see, e.g., Eriugena, *De div.* III, 33 [717–719]. See, e.g., Macrobius, *In Somn.* I, 20, 20, p. 82.
184   Cf. Martianus, *De nupt.* VI, 583–584, p. 205.
185   Richer, *Hist.* III, 50, p. 196.
186   Cf. Isidore, *Etym.* III, 34, 1.
187   Cf. Bede, *De nat. rerum* 5, pp. 196–197; or Isidore, *Etym.* III, 33, 1–2; or III, 36–37, 1.

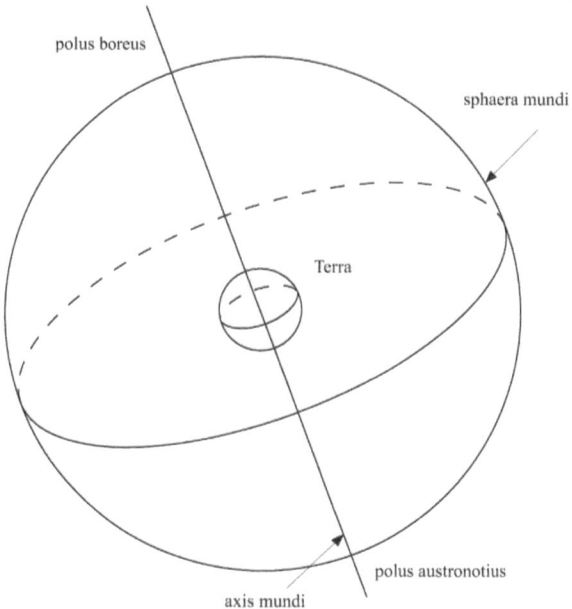

Fig. 12 – Celestial sphere

Richer stresses that through the location (*collocatum*) of the horizon, Gerbert could present to his disciples those constellations which were closer to the north pole as well as the constellations closer to the south pole. It is unclear which stars Richer has in mind. His term *signa* most likely refers to the so-called zodiacal constellations or signs; however, it cannot be ruled out that even the constellations of the southern sky were recorded there.[188] Nevertheless, preserved star tables from Gerbert's time which were available in the Latin West exclusively feature the stars of zodiacal constellations necessary for construction and use of an astrolabe.[189]

---

188 Cf. Lindgren, *Gerbert von Aurillac...*, p. 29. Regarding southern constellations, see, e.g., Martianus, *De nupt.* VIII, 838, pp. 315–316. Hyginus represents another possible direct source.
189 See, e.g., Hermann, *De mens. astrol.* 6, pp. 208–210; Ascelin, *Comp. astrol.*, pp. 350–351.

The horizon was a part of Gerbert's globe and, according to Richer, it could divide visible stars from those which could not be seen.[190] Contemporary authors define the horizon as a circle on the world sphere which delineates the visibility of our sky, i.e., all that we can see in it.[191] The celestial horizon divides the celestial sphere into two hemispheres[192] encompassing different parts of the celestial sphere at various places on the Earth. Therefore, Gerbert could set the horizon on his globe according to the location of the observer.

The horizon defines the visible part of the celestial vault and, concurrently, sets essential navigation points for an observer. All observable stars, at a given time in a given place, appear in the sky at a location, or they gradually rise over the horizon (i.e., they appear and we can see them). In the course of a night, stars gradually rise in the night sky, they reach their highest point precisely in the middle between sunset and sunrise, and then they start to descend or disappear below the horizon.[193] A similar process happens during the day – the Sun rises in the sky after sunset, reaches its peak at midday, and then descends until it sets. Therefore, the horizon represents a basic stable circle that delineates the point of observation.[194] The point that is set perpendicularly above an observer is called the zenith. On the opposite side of the world sphere, under the Earth, we may find the so-called nadir, which can never be seen. All stars culminate (i.e., they are at the highest point in the sky) on a circle that goes twice through the horizon (north and south), through both poles of the world sphere, zenith, and nadir (see Fig. 13). This circle is called the astronomical meridian (*meridianus*).[195]

---

190 Richer, *Hist.* III, 50, p. 196.
191 Abbo, *De spere*, p. 120.
192 Isidore, *Etym.* III, 43, 1.
193 Cf. Cassiodore, *Inst.* II, 7, 2, p. 154.
194 Cf. Macrobius, *In Somn.* I, 15, 17, p. 63; or Cassiodore, *Inst.* II, 7, 2, p. 154.
195 See e.g. Macrobius, *In Somn.* I, 15, 16, p. 63.

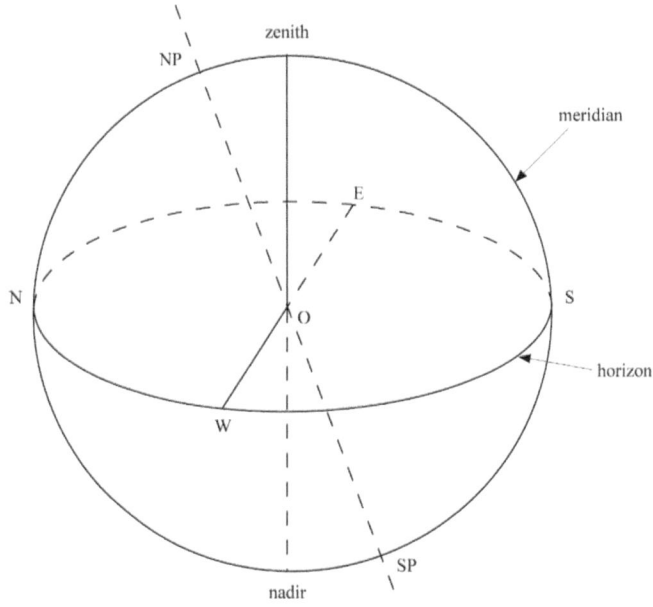

Fig. 13 – Horizon and astronomical meridian
(Abbreviations: O – observation point; N – north; S – south; W – west; E – east; NP – northern pole of the world sphere; SP – southern pole of the world sphere.)

Through the rotation of the globe, happenings in the sky might be simulated. It comprehensively represents the orbits of stars and constellations, since stars follow periodical circles set by the rotation of the world sphere.[196] Since the axis of the world sphere does not share the same tilt as the line connecting the zenith and nadir, it appears to us that some stars in the night sky never set. For an observer situated in the northern hemisphere, these are the stars closest to the north pole of the world sphere – thus called circumpolar or curvilinear constellations – e.g., Ursa Minor and Ursa Major. Other constellations never rise (stars and constellations closest to the southern pole of the world sphere) and we may never see them in the northern hemisphere. Other stars and constellations both rise and set; in these cases, the circle they follow is partially above the horizon and partially below the horizon.[197] Once more, it holds that in the

---

196 Cf. Martianus, *De nupt.* VIII, 815, pp. 309–310.
197 Cf. ibid. VIII, 836, p. 316.

northern hemisphere, we may see the more significant portion of the orbit of those stars which are closer to the north pole and vice versa (see Fig. 14).

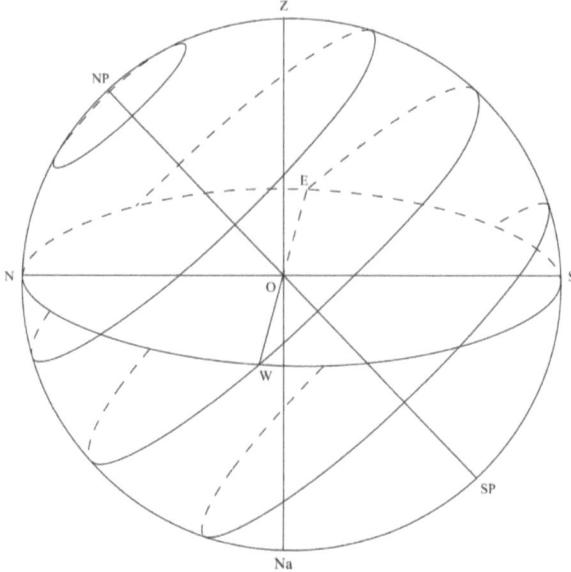

Fig. 14 – Orbits of stars observable over the horizon
(Abbreviations: O – observation point; N – north; S – south; W – west; E – east; NP – northern pole of the world sphere; SP – southern pole of the world sphere; Z – zenith; Na – nadir.)

The stars are visible only at night because the sunlight prevents us from seeing stars during the day and, according to the different lengths of day and night throughout the year, the visibility of various stars and constellations also changes. Seemingly for this reason, Gerbert marked the horizon of his globe to show the rising and setting of stars,[198] enabling him to use this aid anytime and anywhere.[199]

---

[198] For more on these marks on similar globe horizons used in the Muslim intellectual environment, see, e.g., Savage-Smith, E., *Islamicate Celestial Globes: Their History, Construction, and Use*. Washington: Smithsonian Institution Press, 1985, p. 71.
[199] Richer, *Hist.* III, 50, p. 196.

Here, Richer clearly mentions the motivation behind the modification of the celestial globe – it was intended to provide a better understanding of celestial movement. Gerbert's intention was not only to comprehensively demonstrate star orbits to his students but also to allow for an easier understanding of the fact that partially different stars could be seen from various places on the Earth, and that the night sky changes throughout the year, whilst he personally operated the settings of the globe, rising and setting the stars.[200]

Such practical application of the globe is very surprising, since night sky observations and horizon modifications according to the place of observation were unknown in contemporary Latin Europe. Although questions remain regarding Gerbert's knowledge of the astrolabe and other knowledge he attained during his studies on the Iberian Peninsula, this adjustable horizon seems the most conclusive proof of Arabic astronomical influence on Gerbert's own (not exclusively) pedagogical practice.[201] There was a relatively high demand for the celestial globe among Gerbert's disciples, as evidenced by several requests by Remigius of Trier for a model of the Earth sphere, addressed directly to Gerbert.[202] One of Gerbert's responses even provides multiple specific details concerning the creation of this aid. In accordance with Richer's testimony, we may infer that Gerbert made his pedagogical tools himself (or had them made), and that the globe was made from wood. First, a smooth sphere (*spera*) was prepared and covered with horsehide (*equinum corium*). Celestial circles, poles, stars, and constellations were drawn on the surface – all in colour (*color*), and, lastly, the horizon was added to the model of the world sphere.[203]

Gerbert conducted regular observations of the night sky to achieve the proper settings for his teaching aid, and he could set the horizon according to the rising and setting of stars. These observations served only to achieve conformity between the model and the visible sky. For the purpose of making astronomical observations, a more significant role was played by another tool used by Gerbert, according to Richer's *History*.

If an early medieval observer of the night sky wanted to record the orbits of stars or planets, he needed more than just the poles and axis of the Earth, and the horizon, zenith, and meridian. Three reference systems were used for

---

200 Ibid.
201 Cf. Zuccato, "Gerbert of Aurillac…," pp. 758–761. For more details on globes with an adjustable horizon or meridian in the medieval Muslim intellectual environment, see, e.g., Savage-Smith, *Islamicate Celestial Globes…*, pp. 68–74.
202 Gerbert, *Ep.* 134, p. 162; or 148, p. 175.
203 Idem, *Ep.* 148, p. 175.

astronomical practice, through which it was possible to determine the location of individual stars. Among them were the aforementioned coordinates of horizon and meridian; next, there was the system of so-called parallel circles (*aequistances*) perpendicular to the world axis (see Fig. 15). To demonstrate these parallel circles (which, as Richer stresses, are exclusively our auxiliary coordinates – nothing in the sky corresponds to them), we may use Gerbert's second tool.[204] Besides Richer's relatively brief description of the manufacture and function of this tool, we also possess a detailed manual for its preparation written by Gerbert himself, who included it in his letter to Constantine of Fleury.

A necessary component of Gerbert's device showing the basic parallels of the

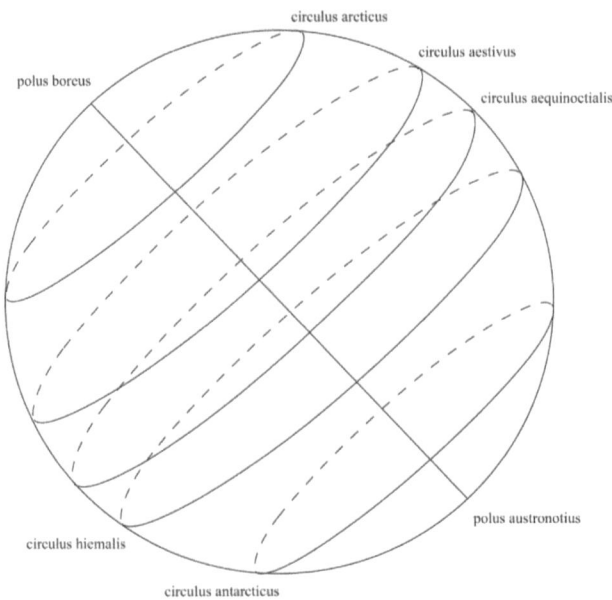

Fig. 15 – Five parallel circles of the world sphere

globe was an observational tube called a *fistula*.[205] It was a hollow tube similar to organ pipes, from which it differed by an equal thickness along its entire body (*aequalis grossitudo*), so that the observation was not hindered by a potential

---

204 Richer, *Hist.* III, 51, p. 196.
205 Gerbert, *De sphaera* 2–3, pp. 27–28; Richer, *Hist.* III, 51, p. 196.

narrowing of the tube.[206] The *fistula* seemingly served to make observation easier through the possibility of aiming it at a specific object and then observing it without risk of confusing it with another (e.g., a different star) – therefore, it served primarily as a tool for ensuring the correct focus.

It is unclear whether the *fistula* was used exclusively for Gerbert's device or whether it simultaneously served as a separate instrument that could have been used by any observer, independent of its role as a teaching aid. A separate use of an observational tube is presupposed by, for example, N. Bubnov, based on Gerbert's remark about *unam fistulam* that helped to identify the northern pole of the world sphere.[207] The same interpretation is supported by Thietmar of Merseburg's mention of Gerbert observing a polar star (*stella nautarum*) via this tube.[208] However, the listed texts do not provide entirely conclusive evidence of this.[209]

And how did Gerbert represent the aforementioned parallels? Richer reports that Gerbert began by cutting a sphere into a hemisphere and then installing the *fistula* along the cut, with the ends of the tube determining the northern and the southern pole of the world sphere. Subsequently, he marked 30 same-length segments across the hemisphere. At the sixth mark, counting from the north pole, he drew a semicircle perpendicular to the axis of the world sphere (i.e., the tube, *fistula*); thus, creating the northern polar circle (*linea arctici*). Five segments further, he drew a second semicircle in the same manner, so marking the Tropic of Cancer (*circulus aestivus*, i.e., the circle of the summer solstice). At the fifteenth segment (i.e., exactly in the middle of the meridian), he marked the equator (*rotunditas aequinoctialis*, i.e., an equinox). Then he proceeded from the southern pole and marked the Antarctic circle and the Tropic of Capricorn.[210]

Gerbert's own instructions to Constantine for the construction of the hemisphere (sphere) are similar: divide the circumference of the sphere into 60 segments; then set the pole anywhere on the line with segments and place one end of a pair of compasses there; next, count six segments and make a circle that delineates 12 points lying on the circumference of the circle. Then, without changing the location of the pole and the anchoring of the compasses, fix the second end of the compasses on the eleventh point, creating 22 segments in total; next, set the anchor of the compasses on the fifteenth mark and draw a third circle,

---

206 Gerbert, *De sphaera* 2, p. 27.
207 Ibid., p. 28 (n. 15).
208 Thietmar, *Chron.* VI, 100, p. 392.
209 Cf., for example, *The Letters of Gerbert*, p. 39 (n. 6).
210 Richer, *Hist.* III, 51, p. 196.

including 30 segments; finally, move the first end of the compasses to the point exactly opposite the original anchoring point and make two more circles in the same manner (i.e., at the sixth and eleventh segments). Thus, all five parallels of the world sphere are created (see Fig. 16).[211]

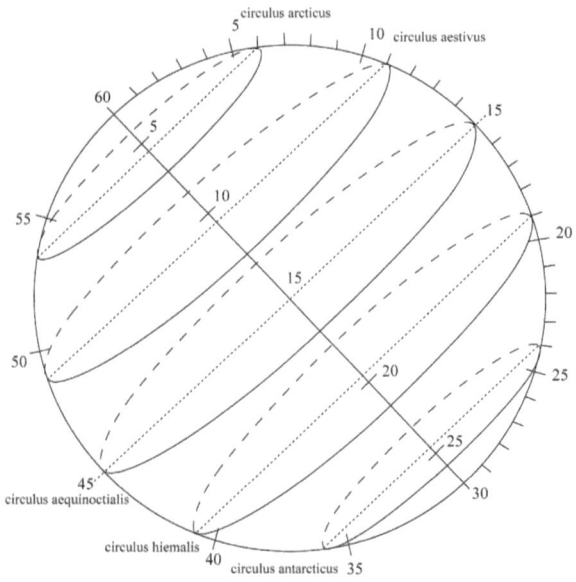

Fig. 16 – Gerbert's placement of the five parallels of the world sphere according to the letter addressed to Constantine *De sphaera*

These five circles of the world sphere[212] – i.e., the northern (*arctica*) and the southern (*antarctica*) polar circles, summer (*aestival*) and winter (*hiemalis*, or eventually *brumalis*, i.e., short-day) tropics, and the equatorial circle (*aequinoctialis*, i.e., the equator) – are marked on the sphere (hemisphere) relatively accurately by Gerbert only in three cases. These cases include the equator and both tropics, which are set at 24° (contemporary measurements approximate

---

211 Gerbert, *De sphaera* 1, pp. 25–27.
212 Cf., for example, Abbo, *De circ.*, pp. 672–673; Bede, *De nat. rerum* 9, pp. 199–200, or Isidore, *Etym.* III, 44, 1–2; etc.

it at 23° 27') since they are placed at the eleventh segment of 15 between the pole and the equator. One segment delineates 6°, which means 24° from the equator.

Gerbert placed polar circles on the ninth segment from the equator, i.e., 54°; whereas today the usually accepted value is approximately 66° 33'. The reason for this choice of location appears to be the text *De Astronomia*[213] by Hyginus, in which the placement of five circles of the world sphere is mentioned in the same context and with the same values.[214] Hyginus's value corresponds to the height of the pole on Rhodes (located at 36° of northern latitude) and follows an earlier tradition (Geminus of Rhodes) which understood the polar circles differently than we do today. The Arctic and Antarctic circle always delineated the visible and never visible part of the celestial sphere, which was dependent on the observation point (horizon), with the value 54° corresponding to Rhodes.[215] Martianus Capella localises these circles similarly, albeit with different delineations.[216]

Nevertheless, let us go back to Gerbert's construction of the given tool. The sphere with circles must be split in half along the circumference, which is used, together with the compasses, to draw all five circles. Thus, two hemispheres are created. It is necessary to hollow them out and drill two holes at both places where the compasses were anchored during the drawing of the tropics, polar circles, and the equator. Two observational tubes (*fistulae*) are installed in each hole (both fit in such a way that it is possible to look through them as one) and their stability is fixed by an iron semicircle (*semicirculus ferreus*), which is made and perforated in the same way as the hemisphere (see Fig. 17).[217]

---

213 E.g., Hyginus' text was available to Constantine or Abbo in Fleury – see Van de Vyver, "Les œuvres inédites…," p. 146.
214 Hyginus, *De astron.* I, 7, pp. 7–8.
215 For more details, see, e.g., Savage-Smith, *Islamicate Celestial Globes…*, pp. 6–8.
216 Martianus, *De nupt.* VIII, 827, p. 312.
217 Gerbert, *De sphaera* 2, p. 27.

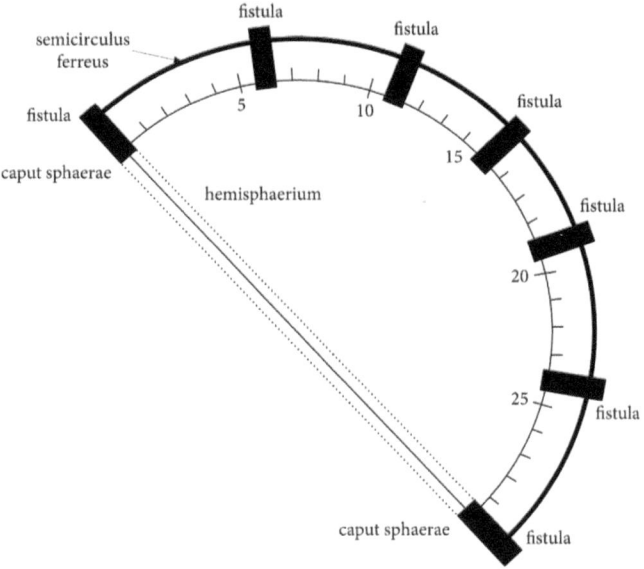

Fig. 17 – Gerbert's observational hemisphere, including an iron semicircle for fixing seven observational tubes

Then, it is necessary to place the hemisphere under the night sky in order to set it correctly, using the two polar tubes. It is sufficientr to locate the star closest to the northern pole of the world sphere, i.e., the Pole Star (*stella polum boreum*) and if it is visible for the entire night, (i.e., it does not change its location) then Gerbert's observational hemisphere is set correctly in accordance to the axis and poles of the world.[218] Observers now have seven tubes at their disposal through which they can view: both poles (*polus boreus, polus antarcticus*) through the first and seventh tube or through the seventh and the first; both polar circles (*circulus arcticus, circulus antarcticus*) through the second and sixth tube; both tropics (*circulus aestivus, circulus hiemalis*) with the aid of the third and fifth *fistula*; and the equator (*circulus aequnoctialis*) with the fourth *fistula* (understandably, the southern polar stars will not be visible since they are always below the horizon in the northern hemisphere; if anybody looked through both polar

---

218 Ibid. 3, p. 28.

tubes in the direction of the southern pole, they would see only the ground, as noted by Gerbert).[219]

It is beyond doubt that anybody could use this observational hemisphere and see where a specific star or a concrete constellation was located at any particular moment and place it in accordance with its equatorial coordinates. However, Richer of Reims, while describing this teaching-observational aid, adds that following the proper positioning (and rotation) of the hemisphere, even inexperienced users (*inexperti*) could observe the circular motion of celestial objects they had not known before and discover and remember these circles.[220] It seems that Gerbert did not create this instrument only as an illustrative aid for the demonstration of the five circles of the world sphere, but also as a tool that allowed for the observation and projection of astronomical processes via parallel coordinates. Therefore, we might say that Gerbert was indirectly encouraging astronomical observations. Further use of this observational hemisphere will be expanded upon later when we move our attention to Gerbert's timekeeping activities.

Another contemporary coordinate system for astronomy, which could be used for recording the orbits of stars and, especially, planets, is the so-called ecliptic, i.e., the Sun's path through the constellations.[221] The Sun does not move parallel to the circles of the world sphere, but its movement occurs in a plane inclined at an angle of approximately 23° 27' to the equator. This movement goes in the opposite direction to the daily rotation of the world sphere. Within a year, the Sun traverses its entire orbit and returns to the same position relative to the stars. The annual motion of the Sun is, therefore, delineated by the constellations through which it gradually passes (consequently, we cannot observe them at the given time). The belt of constellations in the sky through which the ecliptic passes is called the zodiac, in which every constellation (sign, *signum*) represents a twelfth of a circle, i.e., 30° (see Fig. 18).[222]

---

219 Ibid.
220 Richer, *Hist.* III, 51, p. 196.
221 Cf., e.g., Isidore, *Etym.* III, 50, 1.
222 Cf., for instance, Abbo, *De spere*, pp. 120–121; Martianus, *De nupt.* VIII, 834–835, pp. 314–315; or Bede, *De nat. rerum* 16–17, pp. 207–210.

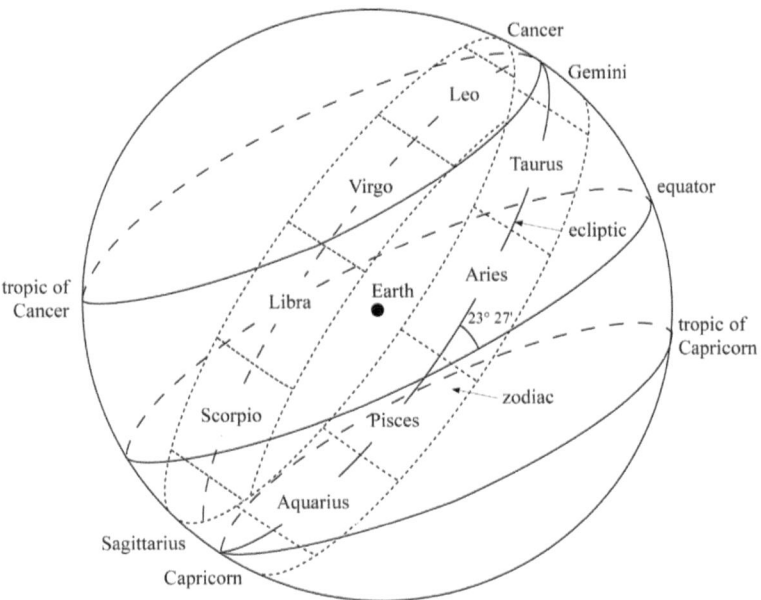

Fig. 18 – Ecliptic, zodiac, and zodiacal signs
(The angle of the ecliptic is significantly higher in this picture to make it more comprehensible.)

Early medieval scholars were fully aware of the difference between planets (*planetae*), stars, and constellations (*stellae, sidera*).[223] While the stars are fixed (*fixus*) to the world sphere, planets are so-called wandering (*errantes*) stars since their movement is different from the regular rotation of the world sphere as they, like the Sun, travel through the zodiac. As a result, for a terrestrial observer, they perform irregular (*anomalus*), e.g., retrograde (*retrogradus*) movements, or they sometimes stop (*stationarius*) in their orbits.[224] The motions of these cosmic objects – usually listed as seven: besides the Sun (*Sol*) and the Moon (*Luna*), there are the five planets, i.e., Mercury (*Mercurius, Stilbon*), Venus (*Venus, Phosphorus*), Mars (*Mars, Pyroeis*), Jupiter (*Iuppiter, Phaeton*), and Saturn (*Saturnus, Phaenon*)[225] – are tied in different ways to the ecliptic, and were

---

223 See Martianus, *De nupt.* VIII, 850–851, p. 322.
224 Cf. Bede, *De nat. rerum* 12, pp. 203–204; see also Isidore, *Etymologiae* III, 64, 1; or Cassiodore, *Inst.* II, 7, 2.
225 Cf., e.g., Martianus, *De nupt.* VIII, 851, p. 322; or Bede, *De nat. rerum* 13, pp. 204–205.

variously explained and recorded by early medieval authors.[226] For the very purpose of depicting the relationship of the ecliptic, the world sphere, the planets, and the Moon to the Sun, Gerbert constructed his third pedagogical aid – the armillary sphere with an ecliptic. First, he connected two circles called *incidentes* in Latin and colures (*coluri*) in ancient Greek and placed the two world poles (*poli*) upon them.[227]

The colures are imaginary circles which run through both poles of the world sphere,[228] equinoctial points, and solstitial points. The equinoctial points are the intersections of the yearly path of the Sun (ecliptic) with the world equator, and both day and night are of the same length at this time, i.e., equinox. Parallel to this, the solstitial points mark the northernmost and southernmost places in which the ecliptic diverges from the equator, i.e., it reaches the tropics of Cancer (summer) and Capricorn (winter). Early medieval thinkers considered the Sun to be the cause of the difference in lengths of day and night throughout the year.[229] Therefore, its orbit determines seasons, which are delineated by the equinoctial points (beginnings of spring and autumn) and tropical points, i.e., the solstices (beginnings of summer and winter). Two colures correspond to these points; the first is called the equinoctial colure and the second is the solstitial colure. The former runs through the northern and the southern world pole and through the points of the autumnal and spring equinoxes, while the latter connects both poles and both solstitial points – summer and winter.[230] For illustration, see Fig. 19.

---

226 For more details, see, e.g., D. Špelda, *Astronomie ve středověku*. Ostrava: Montanex, 2008, pp. 78–85.
227 Richer, *Hist.* III, 52, p. 197.
228 Cf. Abbo, *De spere*, p. 120.
229 See, for instance, Isidore, *Etym.* III, 51, 1–2.
230 Cf., e.g., Martianus, *De nupt.* VIII, 832–833, p. 314.

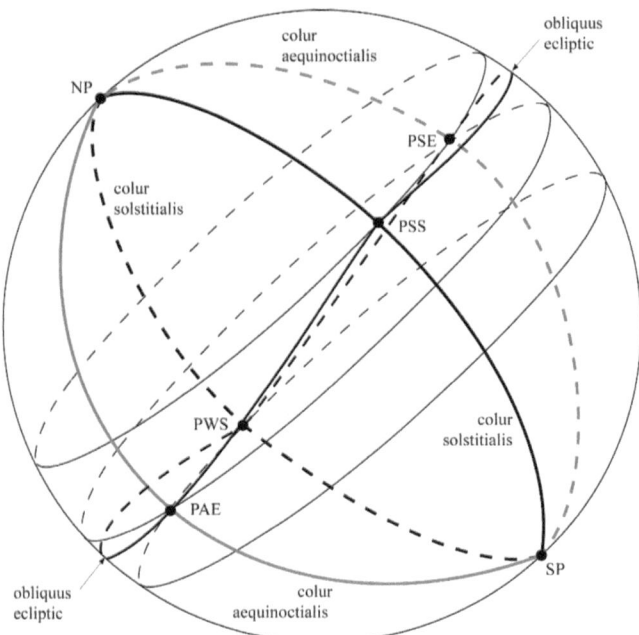

Fig. 19 – The colure of the equinox (*colur aequinoctialis*) and the colure of the solstice (*colur solstitialis*)

(The colure of the equinox runs through the north pole of the world sphere (NP), the point of the spring equinox (PSE), the south pole of the world sphere (SP), and the point of the autumn equinox (PAE); the colure of the solstice runs through the north pole (NP), the point of the summer solstice (PSS), the south pole (SP), and the point of the winter solstice (PWS).)

When Gerbert connected both colures, producing two spherical circumferences that divided the imaginary sphere into four equivalent quarters, he added five parallel circles to the world sphere – i.e., the world equator, the tropics of Cancer and Capricorn, the northern polar circle, and the southern polar circle, placing them exactly as he had on the observational hemisphere. Furthermore, he installed the zodiac at the proper angle to these circles. The last and the most challenging task was to attach the remaining circles representing the movement of the wandering stars (*errantes*), i.e., planets, around the ecliptic (*obliquus*).[231]

---

[231] Richer, *Hist.* III, 52, p. 197.

Gerbert, thus, succeeded in constructing a classical armillary sphere that showed the motions of the planets in relation to the Sun and its orbit. Richer explains the usefulness of this pedagogical aid: it can comprehensively present the apses (*absides*) of the planets, their movable positions relative to the Sun, and their relative distances from each other.[232] There is no doubt about what Gerbert's aid offered concerning the heights and relative positions of the planets. However, the apses are slightly more shadowy. Traditionally, apses refer to the furthest and the nearest points lying on an orbit of a specific celestial body relative to some concrete point (e.g., the Earth) located inside this orbit trajectory. For example, Pliny provided a clear exposition of this concerning the Earth.[233] Furthermore, in his encyclopaedia, Martianus Capella mentions the eccentric (*eccentros*) nature pertaining to the Sun, the Moon, and the planets. Thus, he stresses that these objects might have circular orbits; however, the centres of their orbits are not identical with the centre of the world sphere (i.e., the Earth); in fact they orbit along a circular path with its individual centre located elsewhere, away from the Earth sphere.[234] Whilst, for example, Abbo of Fleury delineates an apse as the position (*locus*) of a planet in the zodiac during the original state (*in principio*), i.e., during the creation of the universe.[235] A concordant interpretation of the planet's positions during the world's birth (*genitura mundi*) was provided by Macrobius, who is the apparent source of Abbo's knowledge.[236] Gerbert, as mentioned before, undoubtedly knew both Macrobius and Pliny. From Richer's description, it is not clear which definition of apses Gerbert subscribed to. Richer might have decided not to discuss other uses of this tool.[237] In Gerbert's texts, there is no mention of apses and it might be presumed that he subscribed to a theory in line with Martianus Capella – that planet orbits are eccentric. Therefore, apses represented the furthest place from and the nearest place to the Earth relative to the eccentric orbit trajectory of the planets.[238]

The final model, to which Richer gives the least attention, was a second armillary sphere. Unlike the first one, metal and iron circles with symbols for zodiacal signs (*figuras signorum*) were added outside the basic circles (apparently,

---

232 Ibid.
233 Pliny, *Nat. hist.* II, 64, p. 147.
234 Martianus, *De nupt.* VIII, 855, p. 314; or VIII, 849, p. 321.
235 Abbo, *De spere*, p. 120.
236 Macrobius, *In Somn.* I, 21, 24, p. 89.
237 Richer, *Hist.* III, 52, p. 197.
238 Cf., for example, Eastwood, B. S., "Astronomy in Christian Latin Europe c. 500–c. 1150." *Journal for the History of Astronomy* 28 (1997), p. 253.

both colures and five world circles).²³⁹ Whilst the sphere, which represented the motion of planets in relation to the ecliptic, seems to primarily possess an illustrative function and introduced planetary orbits to students, this tool (*machina*) was clearly meant for operation under the night sky. It is evidenced by the recurring inclusion of an observational tube into the armillary sphere (here, the *fistula* represents the axis of the world sphere and runs through the intersections of both colures, i.e., the north and the south pole).²⁴⁰

Therefore, the final pedagogical tool mentioned by Richer was meant to be taken by a student, placed under the night sky, set according to the Pole Star (using a *fistula* and the Pole Star), and, thus, used to determine the horizon of the user's observational position. Consequently, it was enough to know the location of one star of a constellation, which was captured on the iron circles on the model's exterior, and the user was able to find the locations of the remaining stars. Richer mentions that this was the case even if the student knew only a little about astronomy and was not accompanied by anybody versed in the science.²⁴¹

As described by Richer and by Gerbert himself in his letter to Constantine of Fleury, these four astronomical tools created by Gerbert clearly show that their primary purpose was to introduce students to the most important findings of astronomical art as known in the contemporary Christian part of Europe. They embodied knowledge of the world sphere, its poles and axis, horizon, meridian, the purpose and construction of the five circles of the world sphere, the relative tilt of the zodiac, the ecliptic, and the difference between planets and stars, etc. Gerbert's tools represented a comprehensive and illustrative introduction to the basics of geocentric astronomy.

While the armillary sphere with the internal zodiac apparently served for teaching planetary motions (including those of the Sun and the Moon), and was used directly during lessons, the globe of the world sphere, as well as the model of the sphere with external constellations, were meant for utilisation under the open sky, and incorporated corrective elements for proper setting (a movable horizon with markings of rising and setting stars, or *fistula* for discovering the world pole). Gerbert obviously remained faithful to his principle of the practical utilisation of gained knowledge; after all, knowledge of the existence of the tropics would serve no purpose if we could not imagine where they were or what we could find in their vicinity.

---

239 Richer, *Hist.* III, 53, p. 197.
240 Ibid.
241 Ibid., p. 198.

A special place among these instruments belonged to the observational hemisphere, to which – in the spirit of Gerbert's letter *De sphaera* – we will turn our attention later. Here, it is enough to say that it could have served a certain role with regard to timekeeping. To this topic, Gerbert devoted another letter, addressed to, the otherwise unknown, brother Adam.

This short text discusses partial theoretical findings, which were fundamental to contemporary timekeeping. It specifically describes changes in the presence of daylight over the horizon during the year in relation to the geographical latitude of a given place, i.e., according to the length of the solstitial day. The letter is accompanied by two tables by means of which Gerbert illustrates the described changes. The text was obviously intended for a highly educated recipient, since Gerbert uses professional terms of contemporary timekeeping, astronomy, and geography without any further explanation. He also does not hesitate to mention two different interpretations of the procedural alternations of the Sun's presence over the horizon during the year, and to explicitly defend one of these interpretations.

Gerbert's letter is apparently a response to Adam's request for clarification of certain principles necessary for timekeeping. We know almost nothing about brother Adam.[242] In the opening of the letter, Gerbert mentions the hardships he has experienced after the death of Adalbero, the Archbishop of Reims.[243] Adalbero's passing to the *intelligibilia*, (which is how Gerbert refers to the death of his relatively close friend, an expression borrowed from Boethius[244]) distracted Gerbert from almost all intellectual projects.[245] Since Adalbero died in January 989, we can, with accuracy, date the letter back to the spring of 989.[246]

---

242 Adam could have been a monk, priest or canon – cf. Gerbert, *Ep.* 153, p. 180 (n. 1) or *The Letters of Gerbert*, p. 190 (n. 1). There is no reason to doubt that he was Gerbert's student, who could have been one of Gerbert's friends and disciples, possibly during the 970s, but most likely in the 980s.

243 For more on Adalbero, archbishop of Reims, cf. Richer, *Hist.* III, 21–23, pp. 181–183; see also Reilly, D. J., "The Bible as Bellwether: Manuscript Bible in the Context of Spiritual, Liturgical and Educational Reform, 1000–1200." In: Poleg, E. – Light, L. (eds.), *Form and Function in the Late Medieval Bible*. Leiden – Boston: Brill, 2013, pp. 13–18; Lake, J., *Richer of Saint-Rémi. The Methods and Mentality of a Tenth-Century Historian*. Washington: Catholic University of America Press, 2013, pp. 22–23; Glenn, J., *Politics and History in the Tenth Century. The Work and World of Richer of Reims*. Cambridge: Cambridge University Press, 2004, pp. 28–46.

244 Boethius, *1 In Isag.* I, 3, p. 8.

245 Gerbert, *Ep.* 153, p. 180.

246 Cf. Gerbert, *Correspondance*, p. 375; or Gerbert, *Lettere*, p. 113.

The very next sentence of the letter provides justification for the hypothesis that Gerbert and Adam were friends. Gerbert states that he wrote the letter in order to make the missing friend (*amicus absens*) present and, as a token of this friendship (*pignus amicitie*), he has chosen several astronomical theses.[247] These theses describe the rising and setting of the Sun (*acessus et recessus solis*) according to the theory that the changes in the duration of the Sun's presence above the horizon are irregular (*inequales*) during the year, rather than according to the interpretation which supposes that, for a specific location, the length of daily sunlight increases (or decreases) regularly (*equales*) every month.[248]

Gerbert subsequently quotes the eighth book of Martianus Capella *De nuptiis Philologiae et Mercurii*,[249] and he reminds us that the increase in daylight, following the winter solstice, proceeds in this way: during the first month, daylight increases by one-twelfth of the difference between the length of the day during winter and summer solstice; in the second month, the daylight increases by one-sixth of the same difference; during the third and fourth month, the growth is one-quarter of the difference between the length of the day during the winter and summer solstices; in the fifth month the increase is again one-sixth; and in the sixth month it is one-twelfth of the same difference.[250] Gerbert does not waste time adding that during the second half of the year, the process is reversed – the length of the day shortens according to the same calculations.

In accordance with this theory, Gerbert sketched the horology of two climates after measuring the length of day for every month in both climates using accurate time intervals (hours). It concerns the climate of the Hellespont (Dardanelles), where the longest day of the year is 15 hours long, and the second *horologium* is meant for a climate whose inhabitants can enjoy daylight during the longest day of the year for a total of 18 hours (see Table 10).[251]

---

247 Gerbert, *Ep.* 153, p. 180.
248 Ibid.
249 Martianus, *De nupt.* VIII, 878, p. 333.
250 Gerbert, *Ep.* 153, p. 180.
251 Ibid.

Table 10 – Gerbert's horological tables (Letter to Adam)

HOROLOGIUM SECUNDUM EOS, QUI DIEM MAXIMUM HABENT HORARUM EQUINOCTIALIUM XVIII.		
Iunius et Iulius	Di. Ho. XVIII	Nox Ho. VI
Maius et Augustus	Di. Ho. XVII	Nox Ho. VII
Aprilis et September	Di. Ho. XV	Nox Ho. VIIII
Martius et October	Di. Ho. XII	Nox Ho. XII
Febroarius et November	Di. Ho. VIIII	Nox Ho. XV
Ianuarius et December	Di. Ho. VI	Nox Ho. XVIII
ITEM HOROLOGIUM ELLESPONTI, UBI DIES MAXIMUS EST HORARUM EQUINOCTIALIUM QUINDECIM.		
Ianuarius et December	Di. Ho. VIIII	Nox Ho. XV
Febroarius et November	Di. Ho. X et semis	Nox Ho. XIII et semis
Martius et October	Di. Ho. XII	Nox Ho. XII
Aprilis et September	Di. Ho. XIII et semis	Nox Ho. X et semis
Maius et Augustus	Di. Ho. XIIII et semis	Nox Ho. VIIII et semis
Iunius et Iulius	Di. Ho. XV	Nox Ho. VIIII

Gerbert processed the tables by this method in order to provide an example (*exemplar*) which could be used by Adam to construct his own horology (*propria horologia*) for any climate. It is enough for Adam to ascertain the length of the solstitial day for a specific place using a clepsydra (*clepsidra*). It is comparatively easy (*facile*) to establish the length of a solstitial day: during a solstice, we must separately mark the amount of water that passed through the clepsydra through the night and through the day, and subsequently, we must convert the sum of these values to twenty-four hours.[252] Both tables are added at the end of the letter.[253]

The orbit of the Sun (or the course of the stars in the night sky) was fundamental to medieval timekeeping based on astronomical phenomena. The focus on time and its counting was dictated by many factors, including the everyday practice or religious duties and needs, for example, regular prayers at monasteries or date assessment of movable feasts, especially Easters, which led to the establishment of an independent interdisciplinary science during the Middle Ages – *computus*, i.e., the *computus paschalis* or *computus eccelsiasticus*.[254]

---

252 Ibid.
253 Ibid., p. 181.
254 For a comprehensive account, see, e.g., Borst, *Computus*...; or Germann, N., *De Temporum Ratione. Quadrivium und Gotteserkenntnis am Beispiel Abbos von Fleury und Hermanns von Reichenau*. Leiden: Brill, 2006.

Although Gerbert did not deal with *computus*, he devoted considerable attention to timekeeping.

Medieval thinkers adopted antique theories about the movement of the Sun, to which two basic motions are ascribed. The Sun and the whole world sphere (*sphaera mundi*), in which the stars (*stellae*) and constellations (*sidera*) are firmly embedded, go around the Earth (from the east to the west) once every 24 hours, and this shift defines day and night. When the Sun is above the horizon, we talk about day, while the absence of daylight is characteristic of night. The Sun and its light distinguish day from night in the same way as light separated day from night during the world's creation.[255] A day is 24 hours long, and one day equals one orbit of the Sun around the Earth.[256]

However, the Sun is not firmly connected to the celestial sphere (contrary to the stars) and it possesses another movement: from west to east. This movement spans over the yearly period and the Sun follows its own circular orbit called an ecliptic (*ecliptica*) orbit, passing through the twelve zodiacal constellations (*zodiacus*).[257] The annual movement of the Sun and its retreat from the celestial equator causes the change of seasons and, of course, the changes in length of daylight (the presence of Sun over the horizon).

Medieval thinkers used two different definitions of an hour (*hora*), derived from these theories about the movements of the Sun. On the one hand, so-called equinoctial or equal (*aequinoctialis, aequalis*) hours were used, with their name derived from equinoctial days during which the Sun is located over the equator and the length of day and night is the same (both are 12 hours). During its regular, and basically equal, daily journey, the Sun travels in a circle (360°), and since a day is 24 hours long, one hour equals the time the Sun needs to cover 15° during its daily movement.[258] This time measurement, the equinoctial (*equinoctialis*) hour, is also used by Gerbert in his letter when he describes the climates.[259]

Nonetheless, the medieval concept of an hour was more often understood as one-twelfth of the time during which the Sun was over the horizon. The day and night thus had the same length (12 hours). Still, the length of day and night changes in relation to the current position of the Sun at the ecliptic, that is, in relation to how long the Sun is above or below the horizon. These hours are

---

255 Bede, *De temp. rat.* 5, p. 283; cf. *Gn* 1, 3–5.
256 Bede, *De temp. rat.* 5, p. 283.
257 See Isidore, *Etym.* III, 50–52; or Martianus, *De nupt.* VIII, 834–835, pp. 314–315.
258 Cf. Gerbert [?], *De util. astrol.* 8, 2–3, p. 132.
259 Gerbert, *Ep.* 153, p. 180.

called temporal or unequal (*temporalis, inaequalis*), since their length changes over the course of a year, depending on the specific date and the geographical latitude.[260]

Both systems of definition imply that medieval scholars divided a day according to four breaking points: noon and midnight (both systems are identical for these two points, since the Sun is at its highest point in the sky at noon, i.e., today we would at precisely 12:00, while at midnight the stars are in the highest place in the sky and we would say that it is precisely 00:00); i.e., the sunrise and the sunset occur at different times during the year, and this is the reason for the different lengths of equal and unequal hours.[261]

Adam apparently must have known of these theories, as Gerbert does not mention the ecliptic or definitions of day and hour in his brief letter, although his statements would not have been comprehensible without this knowledge, since he was using a particular definition of an hour while mentioning different interpretations of the yearly course of the Sun and its timekeeping consequences. Gerbert describes in greater detail the speed at which (in the northern hemisphere) the day is prolonged and night is shortened during the winter and spring months, i.e., daylight declines and the night increases in length over the summer and autumn months. It is clear from Gerbert's text that there were at least two different ways of characterising the chain of changes in the appearance of the Sun over the horizon. Either an equal increase (or decline) of sunlight over a year was presupposed, or these changes were considered unequal.[262]

For a geographical latitude at which the length of day during the summer solstice is 15 hours (as shown in Gerbert's second table), the first interpretation (equal changes) would mean that from June until December, the period of daylight shortens every day by two minutes, i.e., by one hour per month. During the winter solstice, the length of the day is nine hours. This theory, which may have been advocated by the computists of the Carolingian or Ottonian era, fulfils the requirements of regularity, stability, and invariability of events in the sky and their interpretation; however, it corresponds poorly to empirical experience.

---

260 Gerbert [?], *De util. astrol.* 8, 3, p. 132.
261 Time assessment according to equal and unequal hours was also discussed by the contemporary authors of texts about astrolabes and astrolabe constructors, who marked, among others, the curves of unequal hours on their astronomical and timekeeping instruments, which were also modified according to the actual geographical altitude – see below and also, e.g., Gerbert [?], *De util. astrol.* 9–10, p. 133; Hermann, *De mens. astrol.* 3, pp. 206–207; Ascelin, *Comp. astrol.* 3, p. 348; etc.
262 Cf. *De mundi*, c. 883D–884A.

This may have been the main reason why Gerbert himself was more inclined to the second interpretation, and in support of this thesis, he quotes Martianus Capella, who wrote that between both solstices, the duration of the Sun's presence above the horizon changes unequally. According to the aforementioned algorithm, the day in the example of the longest solstitial day (15 hours in the second decade of June) would shorten by 30 minutes on the same day in July; by another 60 minutes in August; by a further 90 minutes in September and October (autumnal equinox); then by another 60 minutes in November; and on the day of the winter solstice in December, the day shortens again by 30 minutes (see Fig. 20).

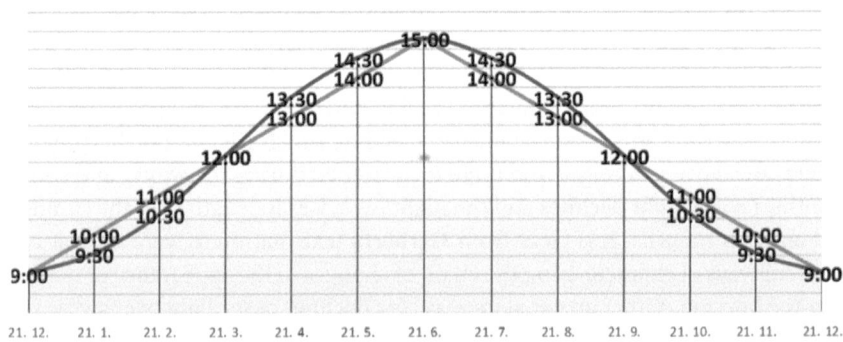

Fig. 20 – Daylight in the climate of Hellespont according to the theory of equal changes and according to the theory of unequal changes

This unequal course of changes in the length of day over the year is accompanied, in Martianus' text, by an interesting reasoning concerning this irregularity. The Sun intersects the equator of the world sphere directly (*directum*) when it goes from south to north in March or when it travels in the opposite direction in September,[263] while during the solstices it must change the direction of its movement, which causes a slowdown since the journey to the north changes into the journey to the south or vice versa. This necessity for a change in direction and the description of the curve, invokes a slowdown of the Sun's movement; therefore, around the equinoctial days, faster changes occur, while changes are slower during the solstitial days. This reasoning may have been seen as plausible by

---

263 Martianus, *De nupt.* VIII, 878, p. 333.

Gerbert. Its application would preserve the regularities in the Sun's movements to a certain extent, which would also correspond better to observations of the sky.

Since Gerbert prepared two horologies for Adam according to this theory – one for the climate of the Hellespont (Dardanelles); the second for a geographical latitude at which the longest day of the year reaches 18 hours – but with no explanation of what the climates (*climate*) are – we can assume that the author of the letter anticipated such geographical knowledge to be known to the addressee.

Adam and Gerbert's educated contemporaries evidently knew about the (not exclusively) contemporary division of the Earth into five basic parts, made with five parallel circles of the celestial sphere (representing the polar circles, tropics, and equator), although here it is applied to the division of the globe (see Fig. 21).[264]

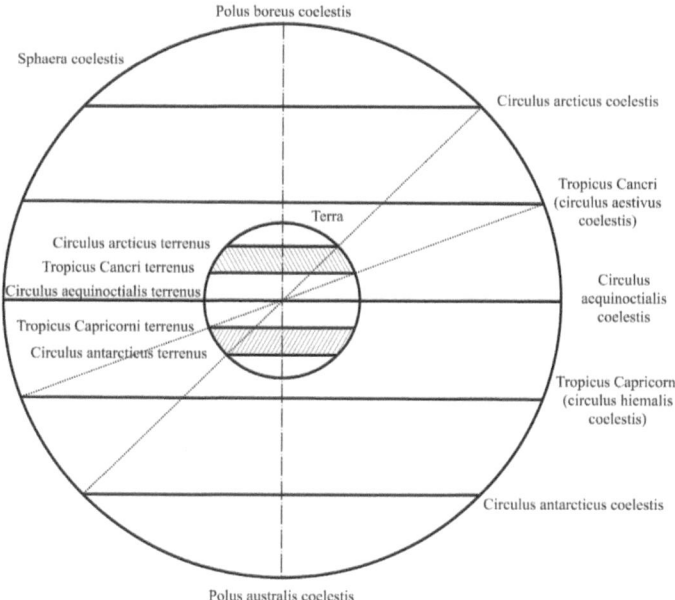

Fig. 21 – Division of the Earth

---

264 Cf. Macrobius, *In Somn.* II, 5, 13–14, p. 112.

According to medieval theories, the territories around the Earth's poles, are demarcated by the polar circles covering the northern and the southern polar areas that were, according to medieval theories, uninhabitable, since the climate was too cold to provide the conditions for a human population. A similar situation can also be found in the part of the Earth located between the tropics, i.e., which enclose the equator on the northern and southern side, where life does not flourish due to the overly hot climate. Two parts of the Earth are left, delimited by two zones stretching between the tropics and the polar circles.

The question of whether the southern part of the Earth was inhabited or not was often tackled during the Early Middle Ages,[265] but in relation to our analysis of Gerbert's letter, we need only focus on the northern hemisphere. The northern inhabited part of the Earth was further subdivided into three continents: the east was formed by Asia, the western part consisted of the northern (Europe) and the southern (Africa) areas. All three continents were divided by the Mediterranean Sea.[266]

Ancient sources introduced the idea of detailed climatic (and time) zones into medieval geographical descriptions, i.e., parallel zones passing through the continents across the same geographical latitude, similar in their climatic conditions, fauna and flora, and human customs.[267] These climates were also demarcated for timekeeping purposes according to the length of the longest day and the shortest night during the year. Scholars usually distinguished between seven zones – i.e., Meroë, Syene (Aswan), Alexandria (Lower Egypt), Rhodes, Hellespont (Dardanelles), Mesopontus (Black Sea), and Borysthenes (the mouth of the river Dnieper),[268] to which other climates were added according to need, or to highlight various extremes or curiosities (such as the mythical Rhypaean mountains or the island Thule in the far north, etc.). By this method, the northern inhabited part of the Earth was divided up into 12 zones usually stretching from

---

265 See, for instance, McCready, W. D., "Isidor, the Antipodeans, and the Shape of the Earth." *Isis* 87/1 (1996), pp. 108–127.
266 Cf. Isidore, *Etym.* XIV, 2; for more details see, e.g., Hiatt, A., "The Map of Macrobius before 1100." *Imago Mundi* 59/2 (2007), pp. 149–176.
267 See, for example, Cassiodore, *Inst.* II, 7, 3; or Isidore, *Etym.* III, 42, 4.
268 See Eratosthenes, *Fragm.* 3A, 18–40, pp. 188 –210; Ptolemy, *Alm.* II, 12, pp. 174–187; Cassiodore, *Inst.* II, 7, 3; Isidore, *Etym.* II, 42, 4; cf., for example, Honigman, E., *Die sieben Klimata und die ΠΟΛΕΙΣ ΕΠΙΣΗΜΟΙ*. Heidelberg: C. Winter, 1929; Stahl, *Roman Science…*; etc.

the Atlantic to India and the Pacific, and including the area from Africa up to the islands in the Arctic Ocean.[269]

In the compendious treatise of Martianus Cappella, the traditional enumeration of seven zones is broadened to eight, while the Black Sea zone (Borysthenes) is omitted and the zones for Rome and the aforementioned Rhypaean mountains are added.[270] Two other marginal northern zones are also mentioned: Britain and the island of Thule.[271] The zone of the Hellespont, which Gerbert introduced into the overview table in his letter, was usually included in the basic enumeration of seven climates and was, among others, characterised by the fact that the longest day spanned 15 hours. However, Capella's description and delimitation of individual geographical zones is somewhat confused – for instance, in the case of the Hellespont, he states that the longest day is 15 hours long, but the shortest night on the same day lasts only eight hours.[272] Gerbert (or Adam) probably used more comprehensive and precise tables, among which the highest clarity is reached in the mentioned passages from Pliny or in the parts of the text by the Venerable Bede, whom Pliny inspired.

However, the standard summaries of climates, available during Gerbert's time, do not include a climate in which the longest day lasts 18 hours, which is approximated by one of Gerbert's horological tables in the letter to Adam. While there are certain references to such a climate during the Early Middle Ages – for instance, the chronicle of the Venerable Bede[273] or Eriugena's commentary to Martianus' *The Marriage of Philology and Mercury*,[274] Gerbert's inclusion of this climate is surprising.

The illustrative and explanatory character of Gerbert's schemes encompassing both climates is the most probable reason for its inclusion in this table. If he had wanted to introduce the theory of unequal change in the duration of sunlight over the course of a year by using specific examples, the most suitable climates for this pedagogical-didactic task would have been those climates where the difference in the length of both solstitial days reaches a value which can be easily divided by the number 12, since the monthly change can be measured as either one-twelfth, one-sixth or one-quarter of the difference between the length of both solstitial days.

---

269 For a more detailed description see Pliny, *Nat. hist.* VI, 33–34(39), 211–220, pp. 517–522; or Bede, *De temp. rat.* 33, pp. 381–386.
270 Martianus, *De nupt.* VIII, 876–877, pp. 331–332.
271 Ibid. VI, 595, p. 209.
272 Ibid. VIII, 877, p. 332.
273 Bede, *Hist. eccles.* I, 1, p. 14.
274 Eriugena, *In Marc.* 296, 5, p. 140.

The climate with a longest day of 18 hours has a shortest day of six hours; therefore, the difference in their length is 12 hours; hence monthly changes can be easily described using whole hours: i.e., 18 hours in June, 17 hours in July, 15 hours in August, 12 hours in September, nine hours in October, seven hours in November, and six hours in December. Similarly, for the climate of the Hellespont, the longest and the shortest day differ by six hours; therefore, the table of this time zone requires only hours and half hours (see Table 11).

Table 11 – Daylight according to Gerbert's two horological tables

	December	January	February	March	April	May	June
		1/12 a	1/6 a	1/4 a	1/4 a	1/6 a	1/12 a	
Climate of Hellespont (a = 6 hours)	9:00	→ 9:30	→ 10:30	→ 12:00	→ 13:30	→ 14:30	→ 15:00
Gerbert's first climate (a = 12 hours)	6:00	→ 7:00	→ 9:00	→ 12:00	→ 15:00	→ 17:00	→ 18:00

(a = difference between the length of the day during winter and summer solstice)

Apparently focusing on explanatory power, Gerbert created this exemplary *horologia* according to which Adam could construct his own horologies, provided he considered the climate in which the given horology was to be applied. The people of the 10th century had several ways of recognizing in which climate (i.e., geographical latitude) they were located.[275] Gerbert offers one simple experimental method, as has been mentioned before. During the solstitial days, a clepsydra can be used. The amount of water that has flowed from sunrise until sunset can be measured and compared with the amount of water that flows from sunset to the following sunrise, and then the given ratio (i.e., the length of day and night) can be converted to 24 hours.[276] By this easy conversion to hours and minutes, anyone interested in creating a *horologium* can easily arrive at the length of the longest day or night.

---

275 For example, an astrolabe can be used for this assessment – see Gerbert [?], *De util. astrol.* 13, 1–2, pp. 134–135.
276 Gerbert, *Ep.* 153, p. 180. A similar method for the assessment of the place of observation (for night observation and time assessment) was described also by Macrobius or Eriugena – see Macrobius, *In Somn.* I, 21, 12–21, pp. 87–88; or Eriugena, *In Marc.* 295, 5, p. 139.

Since we do not know the exact nature of Adam's request to which Gerbert responds, we are left to speculate about the content of the original correspondence between them. The first obscurity resides in the very term *horologium*, which can mean a timekeeping instrument (a clock), but can also refer to the tables concerned with changes in the duration of sunlight over a year for a specific geographical latitude themselves. The diction of the letter itself implies that the second option is more probable, but we cannot entirely reject the first or any other interpretation.

If Gerbert was really writing about the tables, the question of why Adam was interested in their creation would remain. There are multiple possible answers. The table concerning the changes in the duration of sunlight over a year might have been welcomed as an instrument for converting equal and unequal hours. The assessment of climate (geographical latitude) is a necessary condition for the correct set-up of sundials and night clocks; at the same time, it is crucial for the validity of astronomical observations and the correct use of astronomical equipment, including the astrolabe.

We can now transfer our attention to the clockmaking activities of Gerbert. In medieval and early modern primary sources, it is possible to find mention of at least three different clocks that Gerbert is supposed to have constructed during the 10th century, in three different places.[277] The first is the *horologium arte mechanica compositum* in Reims, mentioned by William of Malmesbury. The written records of a clepsydra (*horologium aquatilis sive clepsidra*), which Gerbert is thought to have created in Ravenna, are as late as the 16th century. The oldest reference is offered by Thietmar of Merseburg, who mentions astronomical clocks which Gerbert created and correctly calibrated according to star (*stellae*) observations in Magdeburg.

William of Malmesbury records the clocks in Reims in his brief introduction to Pope Sylvester II (who is incorrectly called John) in *The History of English Kings*.[278] Gerbert is presented – as mentioned above at the beginning of this subchapter – as a man who lusted for glory (*cupiditas gloriae*) and, who therefore, went to Spain (*Hispania*) in order to study astrology (*astrologia*) and other arts (*artes*) with the Saracens (*Saraceni*).[279] He pursued the beneficial (*salubris*) disciplines, such as the traditional arts of the quadrivium, but he also pursued

---

277 Cf. McCluskey, S. C., *Astronomies and Cultures in Early Medieval Europe*. Cambridge: Cambridge University Press, 1998, p. 176.
278 William, *Gesta reg.* II, 167, 1, p. 278.
279 Ibid. II, 167, 1, p. 280.

maleficent (*noxius*) knowledge. He is supposed to have mastered the divinatory art and the art of summoning the souls of the dead.[280] Gerbert allegedly concluded his studies by stealing a precious book (*codex totius artis*). During his escape, he summoned the devil (*diabolus*), with whose help he reached safety and became the devil's vassal (*hominium*).[281]

With the devil's help, he became a successful teacher in Gallia and many important people of the early medieval world were found among his students, including abbots (e.g., Constantine of Fleury and Micy), bishops (e.g., Adalbold of Utrecht, another addressee of Gerbert's mathematical letter), a king (Robert II, son of Hugh Capet), and an emperor (Otto III).[282] These influential people supposedly helped Gerbert secure significant offices in the Church – Robert II was allegedly responsible for Gerbert's appointment as archbishop of Reims,[283] Otto III was thought to have been involved in his appointment as archbishop of Ravenna and as pope.[284]

In other words – everything that Gerbert attempted turned out successfully thanks to the Devil. However, William is aware that similar stories about the dark origins of successful scholars were the mere fabrications of ordinary people (*ficta vulgares*), as suggested by his reference to Boethius, who writes in *Consolation* that he and Philosophy, who taught him, among other things, about the orbits of stars (*sidera*), were accused of being protected by the vilest spirits (*praesidia vilissimorum spirituum*).[285] Nonetheless, William does not doubt Gerbert's profanity (*sacrilegium*), since it is proven by the dreadful circumstances of Gerbert's death.[286]

William's mention of the clocks Gerbert is thought to have created is set in this context. It is listed among the proofs of successful and famous acts that Gerbert achieved with the aid of the Devil: allegedly, it was still possible to

---

280 Ibid. II, 167, 2–3, p. 280.
281 Ibid. II, 167, 4–5, pp. 280–282.
282 Ibid. II, 168, 1, pp. 282–284. For a comprehensive account, see DeMayo, "The Students of Gerbert…," pp. 97–117.
283 Gerbert's struggle for legitimate recognition of his archbishop's office in Reims (expecially 990–995) after the death of Adalbero of Reims resulted in the success of his opponent Arnulf – cf. Carozzi, C., "Gerbert et le concile de St-Basle." In: Tosi (ed.), *Gerberto – scienza, storia e mito…*, pp. 661–676; Dachowski, *First Among Abbots…*, pp. 103–124; or Glenn, *Politics and History…*, pp. 93–109.
284 William, *Gesta reg.* II, 169, 2–3, p. 284. Gerbert was consecrated as Sylvestr II on 2nd April 999, and he stayed in the papal office until his death on 12th May 1003.
285 William, *Gesta reg.* II, 168, 5–6, p. 282; cf. Boethius, *De cons. phil.* I, 4, 39, p. 17.
286 William, *Gesta reg.* II, 168, 6, p. 282; or II, 172, 2, p. 294.

see the hydraulic organ (*organa hydraulica*) and mechanically created clocks (*horologium arte mechanica compositum*) in Reims. The very circumstances in which William mentions these clocks are reasons for doubt. Although Gerbert did indeed operate in Reims for a long time and actually became famous as an expert in philosophy and science (the *quadrivium*), other preserved contemporary reports are silent about any such creations in Reims. Therefore, it is possible to consider William's ascription of the clock's creation to Gerbert as an expression of the chronicler's attempt to attribute Gerbert's successes and extraordinary achievements to the influence of diabolical powers.

William of Malmesbury's claim that Gerbert created clocks *arte mechanica compositum* in Reims is doubtful and, above all, it is difficult to determine which device or tool is supposed to have been used. Translations and interpretations sometimes mention "mechanical clocks,"[287] which is questionable since such clocks are only evidenced from the second half of the 13th century, and more commonly from the 14th century.[288] Another option is to read William's description of Gerbert's clocks as a description of the methods that might be employed in the clock's creation, i.e., they are a piece of craftsmanship and applied knowledge of mechanisms.

The second interpretation might be supported by Richer's *Historia* – the author of this chronicle suggests that Gerbert cooperated with a shield-maker (*scutarius*) to create a calculating tool abacus.[289] Gerbert himself wrote that he had employed craftsmen during the creation of his other astronomical tools – the observation hemisphere or the world sphere.[290] It is possibly not a coincidence that similar words are used in a description of clocks sent by Caliph Harun al-Rashid to Charlemagne and from the description of these clock it is clear that

---

287 See, for example, Dohrn-van Rossum, G., *History of the Hour: Clocks and Modern Temporal Orders*. Chicago – London: University of Chicago Press, 1996, p. 55; or Rollo, D., *Kiss My Relics: Hermaphroditic Fictions of the Middle Ages*. Chicago – London: University of Chicago Press, 2011, pp. 70–71.
288 Cf., e.g., Thorndike, L., "Invention of the Mechanical Clock about 1271 A.D." *Speculum* 16/2 (1941), pp. 242–243; Usher, A. P., *A History of Mechanical Inventions. Revised Edition*. New York: Dover Publications, 1982, pp. 192–196; Hill, D. R., "Clocks and Watches." In: Selin, H. (ed.), *Encyclopaedia of the History of Science, Technology, and Medicine in Non-Western Cultures*, Volume 1: A–K. Berlin – New York: Springer, 2008, p. 153; or Matthews, M. R., *Time for Science Education. How Teaching the History and Philosophy of Pendulum Motion Can Contribute to Science Literacy*. Berlin – New York: Springer, 2000, pp. 56–70.
289 Richer, *Hist.* III, 54, p. 198.
290 Gerbert, *Ep.* 134, p. 162; or 148, p. 175.

it pertains to the method of creation and not to the principle of their function (more details are given below).[291] If the mentioned clocks were not mechanical, but refer to works of craftsmanship and knowledge of the necessary arts, then the same methods may have been used to create the astrolabe or nocturlabe.[292] Indeed, even Thietmar's description of Gerbert's clocks in Magdeburg, which was correctly calibrated by an observation tube (*fistula*), could be interpreted as a clock of this type. If the planisphere astrolabe (an astronomical tool which employed stereographic production to mark the celestial sphere with its circles on a plane and, was, concurrently, marked with the coordinates of the horizon, the so-called almucantars – see Fig. 22) was used, then the said observation tube was an alhidade, which served, for example, for observation of the Sun and bright stars.

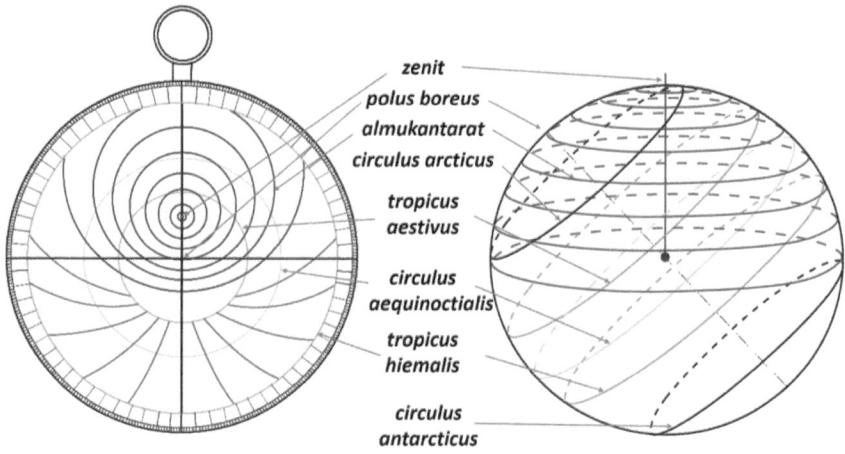

Fig. 22 – A celestial sphere with the Earth in its centre and its image on the front side of an astrolabe

This alternative interpretation may be supported by the fact that Gerbert is traditionally associated with the introduction of the astrolabe in the Latin

---

291 See, for instance, Truitt, E. R., *Medieval Robots. Mechanism, Magic, Nature, and Art*. Philadelphia: University of Pennsylvania Press, 2015, pp. 146–147.
292 Cf. Thompson, R, M. – Winterbottom, M., *William of Malmesbury, The History of the English Kings: General Introduction and Commentary*. Oxford: Clarendon Press 1999, p. 155; or Poulle, E., "Gerbert horloger." In: Guyotjeannin, O. – Poulle, E. (eds.), *Autour de Gerbert d'Aurillac, le pape de l'an mil*. Paris: École des chartes, 1996, p. 366.

Christian West. It was not rare to ascribe the authorship of one of the oldest Latin texts on the astrolabe *De utilitatibus astrolabii* to him, although it is more often supposed today that the text is not Gerbert's own work but that it originates from one of his disciples or colleagues.[293] Authorship notwithstanding, it is clear that Gerbert was more or less involved in the growing interest in the astrolabe, about which other texts are written at the beginning of the 11th century, providing instructions for its construction and use.

In the text *De utilitatibus astrolabii*, the fact that it is a perfect device for timekeeping is stressed in many places. The very first chapter puts astronomical findings, such as ascending and descending constellations (*ortus et occasus siderum*), the stars (*stellae*), and even the Sun (*sol*) in direct relation to the most precise (*certissima*) methods for determining hours, both day and night (*noctes et dies*), and natural and artificial (*naturales sive artificiales*); and also in relation to the determination of the temporal (and geographical) climate (*clima*), which is necessary for timekeeping (*horologium*).[294] Similarly, the text mentions that timekeeping by means of the astrolabe is the most suitable or appropriate (*dignissimum*) method for determining daytime hours (*diurnae horae*).[295]

This understanding of the astrolabe in *De utilitatibus astrolabii* corresponds to the structure of the given treatise. Essentially, the text focuses on time and everything related to timekeeping. First (chapter 3), the method of locating the Sun's current position on the ecliptic is introduced (in terms of degree of zodiac and the appropriate constellation) along with the method of determining the nadir of the Sun's current position, i.e., a degree in the exact opposition to the Sun

---

293 Cf., for example, Bergmann, W., *Innovationen im Quadrivum des 10. und 11. Jahrhunderts. Studien zur Einführung von Astrolab und Abakus im Lateinischen Mittelalter*. Stuttgart: Steiner Verlag, 1985, pp. 66–174; Borst, A., *Astrolab und Klosterreform an der Jahrtausendwende*. Heidelberg: Universitätsverlag Carl Winter, 1989; idem, *The Ordering of Time. From Ancient Computus to the Modern Computer*. Cambridge: Polity Press and Blackwell Publishers, 1993, pp. 54–63; Van de Vyver, "Les plus anciennes Traductions…," pp. 658–691; Stautz, B., "Die früheste bekannte Formgebung der Astrolabien." In: Von Gotstedter, A. (ed.), *Ad radices. Festband zum fünfzigjährigen Bestehen des Instituts für Geschichte der Naturwissenschaften der Johann Wolfgang Goethe-Universität, Frankfurt am Main*. Stuttgart: Steiner Verlag, 1994, pp. 315–328; or monothematic issue of journal *Physis: Rivista internazionale di storia della scienza* 32 (1995) – *The Oldest Latin Astrolabe*, edited by Wesley M. Stevens, Guy Beaujouan and Anthony J. Turner.

294 Gerbert [?], *De util. astrol.* 1, 2, pp. 115–116.

295 Ibid. 5, 4, p. 129. An astrolabe is referred as the most certain (*certissimus*) tool for timekeeping also in ibid. 6, 1, p. 130.

(chapter 4). Then the method of determining day hours, according to the altitude (*altitude*) of the Sun (chapter 5) and night hours according to the altitude of stars (chapter 6) is described. The following chapter deals with marking the temporal (irregular) hours on the astrolabe disk according to the daily movements of Sun (chapter 7) and clarifying the difference between regular and irregular hours (*horae aequinoctiales et inaequales*). As mentioned above, the former expresses 1/24 of the Sun's daily orbit around the Earth, i.e., a shift of the Sun in the sky by 15°, which means that every hour has the same length (and this is the way we define an hour even today). Irregular hours are not of the same length, since the Sun is above (or below) the horizon for a different period of time; however, the presence of the Sun above the horizon is always divided into 12 hours, as is the time of night, when the Sun is below the horizon. During the year, the length of hour changes corresponding to the length of the journey the Sun travels above the horizon – during the summer, this may be, for example, 19°, and one hour lasts 76 regular minutes, while during the night the Sun may travel at 11°, and one hour lasts 44 regular minutes (chapter 8).

In addition, the text *De utilitatibus astrolabii* describes how to determine irregular hours or their parts during the day (chapter 9) or during the night (chapter 10), and goes on to describe the analogy for regular hours (chapters 11–13). The next parts of the text are dedicated to star rise (chapter 14) and to the determining of other data necessary for finding out the current time according to the orbits of celestial objects (chapters 15 and 16). Consequently, the stars of the constellations and their shapes are introduced since they are essential to the operation of the astrolabe (chapter 17), and this is accompanied by a detailed description of time and climate zones (whose delimitation is determined according to the length of the equinoctial day), including their geographical descriptions (chapters 18 and 19). The final part of the text focuses on findings important to gaging whether or not the noon has already taken place during the given day (chapter 20), and goes on to describe timekeeping using the reverse of the astrolabe (chapter 21). In the final sentence, the author clearly states that the astrolabe (*walzagora, id est plana sphaera*) can serve as a clock (*horologium*) in this way.[296]

---

[296] Ibid. 21, p. 147.

This brief overview of the utility of an astrolabe in the given text, which probably came from Gerbert's friends, disciples, or colleagues, implies that the astrolabe was understood primarily as a tool for timekeeping. Provided we have placed the disk with the inscribed round celestial spheres, including almucantars, in correspondence with our current observing position (i.e., provided that the disk corresponds to the climate we are in), then the determination of time by the astrolabe is relatively simple (see Fig. 23):

1. We determine the current position of the Sun in the ecliptic (e.g., with the help of the reverse of the astrolabe, on which the calendar is located, with every day corresponding to a degree of zodiac).
2. On the *rete*, the front side of the astrolabe, we mark the current position of the Sun and its nadir.
3. We measure the height of the Sun above the horizon through the *alhidade* (once again, on the reverse of the astrolabe).
4. We determine whether it is morning or afternoon.
5. On the front side of the astrolabe, we rotate the sign where the current position of the Sun is marked on the *rete*, according to the current measured height of the Sun above the horizon and the corresponding degree of *almucantars* which are drawn on the front side of the astrolabe (if it is afternoon, we put a mark on the left part of the astrolabe and if it is morning, then on the right side of the astrolabe).
6. The nadir of the Sun will show us the current time according to the curves of irregular hours on the front side of the astrolabe.

120 Theoretical Arithmetic

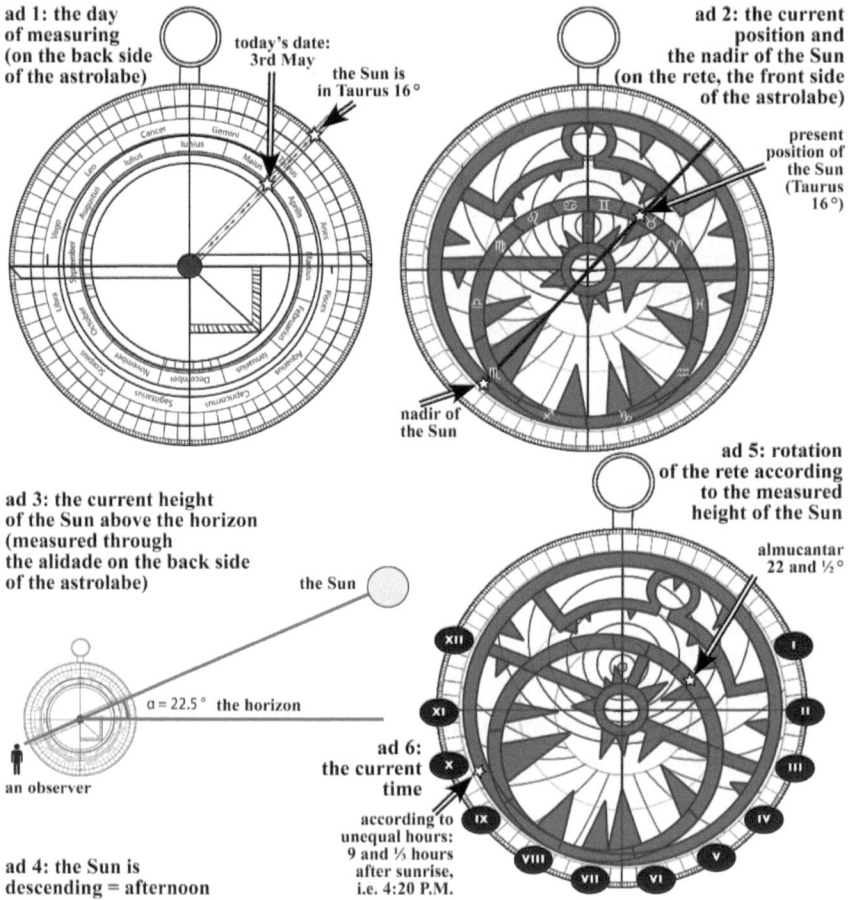

Fig. 23 – Determining the current time using the astrolabe

If we wish to transfer this time to regular hours, then we can simply count degrees on the edge (*limbus*) of the astrolabe from noon, or from sunrise, and then divide the result by 15, i.e., the 15° travelled by the Sun in one regular hour – according to the starting point of our calculation, we arrive at the time expressed by regular hours that have passed since noon or sunrise.

Therefore, we cannot exclude the possibility that Gerbert presented certain information concerning the astrolabe as a timekeeping device in Reims. Although Richer of Reims is silent about Gerbert's astrolabe, at least one passage in Gerbert's correspondence could be interpreted as a reference to timekeeping with the help of the alhidade on the astrolabe.[297] If William's mention of Gerbert's clocks in Reims is to be given any credence, it must refer to his creation of an astrolabe, which he had the necessary knowledge to use as a timekeeping device.

The references to Gerbert's clepsydra in Ravenna do not fare much better. Once again, it is true that Gerbert stayed in Ravenna – during the years 998–999, he held the position of archbishop[298] – however, his time in Ravenna was relatively short, and no sources from his time (or immediately after his death) mention any such device.

One report originates from the end of the 16th century when Benedictine monk Arnold Wyon (Vuion) dedicated a chapter of his extensive work *Lignum vitae* to archbishop Gerbert, who later became Pope Sylvester II. His text focuses almost exclusively on linking Gerbert to the construction of clocks and organs. Following a list of several quotations mentioning Gerbert's creations, he reminds us that similar devices were known even before,[299] and consequently, he mentions the clepsydra (*horologium aquaticum sive clepsydra*) in Ravenna, which was supposedly created by Gerbert during his time as archbishop.[300]

In the 17th century, the Jesuit Augustine Oldoini writes in a very similar manner when he complements the *Vitae et res gestae Pontificum romanorum et S.R.E. Cardinalium* (first published in 1601) of the Dominican scholar Alphons Ciacono. Ciacono sketches a fairly standard brief account of Gerbert's life and his intellectual successes (the clocks in Magdeburg are not neglected) to which he adds narrative passages about Gerbert's pact with the Devil.[301] Oldoini includes further information about Gerbert's life and the legend surrounding his character. In addition to the clocks and organs from Reims, Ravenna's clepsydra (*horologium aquatilis seu clepsydra*) is also mentioned, the formulation of which is nearly identical to that of Wion.[302] However, it is difficult to suppose that Gerbert's clock in Ravenna, basically unknown for six centuries, should suddenly have been rediscovered. None of these reports of the clepsydra can be considered reliable.

---

297 Gerbert, *Regulae*, praef., pp. 7–8.
298 For more details, see, e.g., Vasina, A., "Gerberto arcivescovo di Ravenna." In: Tosi (ed.), *Gerberto – scienza, storia e mito…*, pp. 255–272.
299 Wyon, A., *Lignum vitae* V, pp. 741–743.
300 Ibid., 743: *Horologii uero aquatici, siue Clopsydrae figura, est Rauennae in Herculis regione, quam ipse idem construxit dum illic esset Archiepiscopus.*
301 Ciacono, A. – Oldoini, A., *Vitae pontif. card.*, c. 751A–754D.
302 Ibid., 756A.

Although Gerbert probably did not create the clepsydra during his stay in Ravenna, it is beyond doubt that he knew of such clocks. From Cassiodore,[303] we know that such clocks were used in the Early Middle Ages, and that they were known in other parts of Europe during Gerbert's time.[304] Even the famous clocks (*horologium arte mechanica compositum*) wonderfully (*mirifice*) crafted from brass (*ex auricalcum*) that the caliph of Baghdad Harun al-Rashid sent to Charlemagne in 807 were water timekeeping devices (*clepsidra*).[305]

It is apparent that the clepsydra was well known during the Carolingian and Ottonian eras. Even Gerbert himself writes about it in the horologic letter to brother Adam (analysed above) in which he introduces practical instructions on how to identify the time zone in which an individual is located. It is clear from this letter that he was experienced in using the clepsydra, and that he actively used it for timekeeping. However, it is not possible to assert that he created such a mechanism in Ravenna only from the preserved historical sources.

On the other hand, the case of the clocks Gerbert is supposed to have created in Magdeburg is different. They are referenced by Thietmar of Merseburg, bishop and close friend of Henry II, who became Holy Roman Emperor in the year 1002 after the demise of his cousin Otto III. Thietmar focuses on the deeds of Henry II in his chronicle. At the end of the sixth volume, in one of many digressions (this one preceding the account of the coronation of Holy Roman Emperor Henry II by Pope Benedict VIII in 1014), Thietmar briefly lists Benedict's predecessors in the Holy See since Gregory V (pope in 996–999). The only pope he examines in relative detail is Gerbert, mentioning his illegitimate (*iniustus*) position as archbishop of Reims, extensive education in natural disciplines (*naturales artes*), and unparalleled knowledge of astronomy (*astrorum cursus*), by which he surpassed (*superare*) all of his contemporaries (*contemporales suos*). After he was forced to leave Reims, he moved to the emperor's court, located in Magdeburg at that time.[306] There Gerbert is said to have created clocks (*horologium*) that he correctly calibrated according to the pole star (*stella, dux nautarum*), observed through an observation tube (*fistula*).[307] The brief biography of Gerbert by Thietmar

---

303 Cassiodore, *Inst.* I, 30, 5, pp. 77–78; or idem, *Var.* I, 45, pp. 39–40.
304 See, e.g., *Consuet. Flor.*, p. 42; or Abbo, *In Calc.* III, 37, p. 95; cf. also Evans – Peden, "Natural Science…," pp. 119–120.
305 *Ann. Franc.* 807, pp. 123–124.
306 For more information on Gerbert's relationship with Ottonian imperial dynasty, see, e.g., Lindgren, *Gerbert von Aurillac…*, pp. 69–94; or Zimmermann, H., "Gerbert als kaiserlicher Rat." In: Tosi (ed.), *Gerberto – scienza, storia e mito…*, pp. 235–253.
307 Thietmar, *Chron.* VI, 100, p. 393.

completely omits any references to the Devil (negative legends of Gerbert began to form significantly in later years), and only his knowledge of the liberal arts is mentioned.

The credibility of this report is further supported by the fact that in the 990s (after he had stayed at Quedlinburg Abbey and at the Monastery of St. John the Baptist near Magdeburg) Thietmar was a canon of the cathedral chapter of St. Maurice in Magdeburg.[308] He directly witnessed Gerbert's stay at the emperor's court in the city during the third quarter of the 990s, and it cannot be ruled out that one of the few details about Sylvester II listed in his chronicle is included because Thietmar witnessed Gerbert's creation of it with his own eyes.

Thus, we can consider the *horologium* in Magdeburg to be the work of Gerbert with some degree of confidence. Unfortunately, Thietmar's description is very brief and, therefore, it is very difficult to learn what the device entailed. Essentially, we only know that it was a *horologium* which Gerbert set by observing the pole star through an observation tube (*fistula*).

If we take this *fistula* to be an *alhidade*, then the tool used was an astrolabe, which can suitably serve as a timekeeping device. Thietmar's mention of observations of the night sky (the pole star) could be seen as an attempt by Gerbert to precisely determine the observation area so that the astrolabe disk could be adjusted to correspond to the current climate; thus, the astrolabe could have served as a tool for timekeeping purposes in Magdeburg.

Another possibility is that it entailed an alternative tool for night timekeeping. Indeed, such a machine is ascribed to Pacificus of Verona (who died during the first half of the 9th century), on whose gravestone (although only dating from the 12th century[309]) is written that he had created night clocks (*horologium nocturnum*) never seen before, and that he had devised a method (*argumentum*) of determining the time with their help.[310] This epitaph also adds that Pacificus wrote more than two hundred books, including a horological poem about the celestial sphere.[311] Through interpretation of this poem and preserved

---

308 Ibid. IV, 16, p. 151.
309 Cf. La Rocca, C., "A man for all seasons: Pacificus of Verona and the creation of a local Carolingian past." In: Hen, Y. – Innes, M. (eds.), *The Uses of the Past in the Early Middle Ages*. Cambridge: Cambridge University Press, 2004, pp. 250–257.
310 *Epit. Pacif.*, v. 12–13, p. 655.
311 Ibid. v. 15, p. 655.

illuminations from contemporary manuscripts, we can explain how this clock would have operated.[312]

It consisted of an adjustable observation tube (referred to as *fistula* in contemporary texts[313]) with a fixed stand, connected to a circular disk (*rota*) on which lines of hours were inscribed, as well as the solstitial days (*solstitial*) and equinoctial days (*aequinoctial*), together forming a cross (*crux Christi*). With the help of the observation tube, the highest star closest to the northern pole of the celestial sphere (*polus*) was found. The celestial sphere (*spera caeli*) revolved around (*revolvere*) its axis (*axis*), passing through its poles once in 24 hours (*horae quarter senis*), and the observation tube followed the axis of the world sphere.[314] The machine's location was fixed by adjusting it to the given observation place according to the celestial sphere.

From the north pole to the south, lines could be drawn (*rectae lineae*) which corresponded to the hours on the device's disk (the distance between both equinoctial days marked on the disk was 180°, which corresponded to twelve hours, i.e., the revolving of the celestial sphere by 15° in one hour). Therefore, due to the daily rotation of the celestial sphere, the stars woven into it (*stellae fixae*) followed bigger or smaller circles (*circuli*) – the closer they were, the shorter (*berviores*) their orbits. The observer (*curiosus*) must then find the star in the sky according to whose movement he could determine the time (*computatrix*). The northern circumpolar stars were suitable for the northern hemisphere since they could be observed throughout the entire year. According to the image from the lost manuscript Chartres 214, this could have meant the star α UMi from Ursa Minor, i.e., the tail star from the Lesser Bear constellation, known as the Pole Star nowadays. This star was located at approximately 7° from the northern pole of the celestial sphere in the 9th century and, in contrast to today, was not suitable for determining the northern pole.[315] Nevertheless, its clear visibility and close proximity

---

312 See, for example, Wiesenbach, J., "Pacificus von Verona als Erfinder einer Sternenuhr." In: Butzer, P. L. –Lohrmann, D. (eds.), *Science in Western and Eastern civilization in Carolingian times.* Basel: Birkhäuser Verlag, 1993, pp. 229–250.

313 See, e.g., Avranches, Bibliothèque Municipale, MS 235, fol. 32v; or the manuscript destroyed during the Second World War Chartres, Bibliothèque Municipale, MS 214, fols. 31 or 32 – cf. Michel, H., "Les tubes optiques avant le telescope." *Ciel et terre: Bulletin de la societe beige d'astronomie, de meteorologie et de physique du globe* 70 (1954), p. 177.

314 Pacificus, *Spera*, p. 692.

315 Cf. Wiesenbach, J., "Der Mönch mit dem Sehohr. Die Bedeutung der Miniatur Codex Sangallensis 18, p. 45." *Schweizerische Zeitschrift für Geschichte* 44/4 (1994), pp. 380–382.

to the pole allowed for simple night observations of its orbit for the entire year; therefore, it could have served as a suitable time indicator (*computatrix*). The movement of this star in the night sky moved above the periphery of the disk (*volvens*) of the measuring device, and, hence, the current position of the star in the sky served as a night clock indicator (*horae noctes*) on the disk (see Fig. 24).

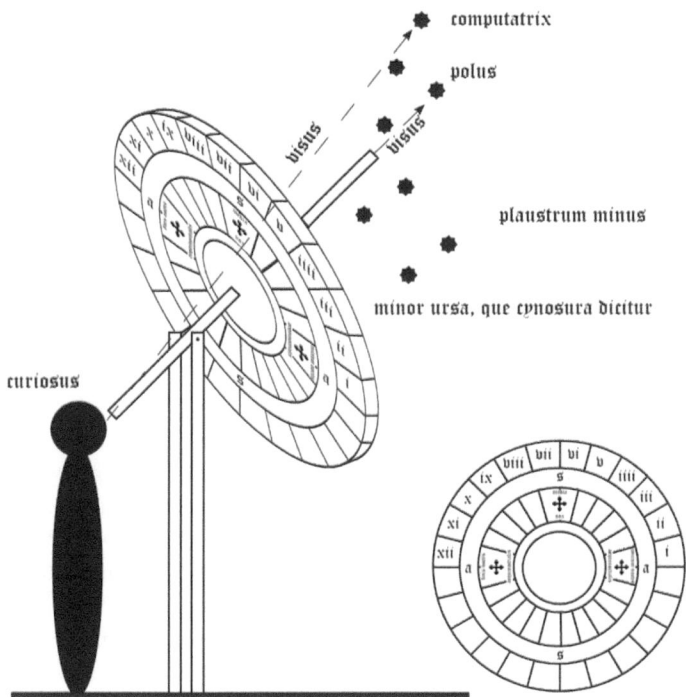

Fig. 24 – Possible appearance of clocks ascribed to Pacificus of Verona

Due to the Sun's annual movements in the ecliptic, and the uneven length of night during the year, the rotary disk needed to be readjusted daily (by an incomplete 1°).[316] The clock user had to check the current rise of the stars to calibrate the night clock, but he was also able to recognise which part of the year it was currently thanks to the equinoctial and solstitial days. According to the

---

316 See ibid., p. 383.

marked lines of individual hours, the time could be determined in regular hours (as implied by the image from St. Gallen, Stiftsbibliothek, Codex Sangallensis 18, 43[45])[317]) or in irregular hours (with regard to the imperfect nature of the illustrative picture, see the image from Rome, Biblioteca Apostolica Vaticana, MS Vat. lat. 644, fol. 76r[318]).[319]

These clocks could very well correspond to what Thietmar of Merseburg describes as Gerbert's creation in Magdeburg. According to Gerbert's letter *De sphaera* and Richer's *History*, Gerbert used observational tubes (*fistulae*) fixed to a hollow hemisphere and their location corresponded to the northern and southern poles of the world sphere, both polar circles, both tropics, and the equator.[320] The tubes passing through the northern and southern poles copied the axis of the celestial sphere, while the remaining five tubes allowed for night sky observations. From both descriptions, it seems that Gerbert built an observation aid in Reims via this method, although it is clear that, with a knowledge of star orbits, the hemisphere could have served for timekeeping as well. Gerbert's *horologium* from Magdeburg may have been similar to Pacificus's night clocks, employing the observation hemisphere he used actively in Reims. Indeed, we might interpret the aforementioned description by William of Gerbert's *arte mechanica compositum* clocks in Reims as referring instead to the creation of the *horologium*; meanwhile, the observation hemisphere and night clocks could have been confused in later tradition, since both devices necessitated knowledge of the astronomical arts for their creation. The very fact that William mentions both of Gerbert's successes in Reims together – i.e., organs and clocks – might have been due to the fact that the tubes (*fistulae*) are called by the same name in Gerbert's texts, although constructed differently – without them, neither organs, nor the clocks which Gerbert reportedly set in Magdeburg could have functioned.

It is beyond doubt that Gerbert dedicated a significant part of his work to timekeeping. His astronomical observations influenced his approach to time. It is clear that he was aware of the clepsydra, which he recommends to brother Adam as a tool for the determination of the geographical latitude (*clima*) in which the given observer is located. Therefore, the clepsydra is introduced specifically as a

---

317 See https://www.e-codices.unifr.ch/en/csg/0018/43/0/Sequence-235 [2019-04-15].
318 See https://digi.vatlib.it/view/MSS_Vat.lat.644 [2019-04-15].
319 Cf. Oestmann, G., "On the History of the Nocturnal." *Bulletin of the Scientific Instrument Society* 69 (2001), pp. 5–6.
320 See above pp. 95–96, cf. Richer, *Hist.* III, 51, p. 196; and Gerbert, *De sphaera* 2–3, pp. 27–28.

tool necessary for obtaining important data for astronomical observation. It was also common during the era for sun clocks to be used for timekeeping.

However, for night timekeeping, it was necessary to use other tools. An astrolabe is an obvious choice, since it served – as evidenced by the text *De utilitatibus astrolabii* ascribed (probably incorrectly) to Gerbert – for timekeeping purposes. Since we lack any substantial evidence concerning Gerbert's active use of the astrolabe, we could speculate that he might have attempted to create a device which may have been a modified version of the tool ascribed to Pacificus of Verona – night clocks that require the setting of the timekeeping device by the celestial sphere and its northern pole for correct functioning. This corresponds to the characterisation of Gerbert's *horologium* by Thietmar of Merseburg, whose account – in contrast to other references to clocks constructed by Gerbert – is comparatively reliable.

These practical applications form a substantial part of Gerbert's mathematical activities – arithmetic, as a vital prerequisite of astronomical observation, thus gains another critical area of applicability. Therefore, Gerbert utilises theoretical arithmetic not only for his own research in the field of arithmetic, but he also applies its findings to issues concerning metaphysics, theology, anthropology, and many others, which are realised not only in the field of arithmetic but also in cooperation with the other disciplines of the quadrivium – geometry, music, and astronomy.

II. PRACTICAL ARITHMETIC

Saltus Gerberti, Geometria, the organ, *De sphaera, horologium,* the astrolabe, the observational hemisphere, etc., are renowned achievements tied to Gerbert's name and they stem from his conception of arithmetic, which innovates on the contemporary understanding of this liberal art. The very application of arithmetic to other quadrivial sciences demonstrates a practical aspect to Gerbert's approach to the science of numbers, which is further expanded upon in the form of practical arithmetic, i.e., (in Plato's aforementioned words) in counting.

Two of Gerbert's "primacies" are often stressed regarding basic arithmetic and numerical operations – the reintroduction of the counting tool known as the abacus (which, under Gerbert, takes the form of a table with inscribed columns of decimal orders), and the first introduction of Western Arabic numerals to the Latin Christian West, which is tied to the new method of numerical value notification using the positional decimal system.

Therefore, the second part of this book will focus predominantly on the abacus and the form in which it was presented by Gerbert to his disciples and friends, including the appearance of Western Arabic numerals and the recording of numerical values by them. In addition, attention will shift to Gerbert's description of a method for making arithmetical calculations on an abacist counting table. In accordance with Gerbert's treatise *Regulae de numerorum abaci rationibus*, we will first look at the rules for multiplication, followed by those for division.

1. *Tabula abaci*, Decimal Positional Notation and *Ghubar* Numbers

At the end of the *Constitution of the Athenians*, Aristotle describes the voting process used during judicial trials. Judges receive both pebbles with a hole and whole pebbles. The former represents a vote in favour of the plaintiff, while the latter expresses a vote in favour of the defendant. According to their conclusions, judges place pebbles representing their stance inside a bronze amphora while the second pebble is thrown into a wooden amphora.[1] When the voting process has been concluded, the pebbles from the bronze amphora (i.e., valid votes) are tipped out onto a special table (ἄβαξ) with holes, into which the individual

---

1 Aristotle, *Ath. pol.* 68, 2–4, pp. 63–64.

pebbles are inserted. The perforated and whole pebbles are inserted into different holes, allowing for easier counting and deciding of the outcome of the vote.[2]

Thus, Aristotle briefly describes an abacus. It is a tool that makes tallying votes more manageable and more comprehensible. The structure and the construction of this table allow for a visual demonstration of which side gained the most votes at the trial. Thus, we can say that it is a simple tool that makes an arithmetical operation easier and, due to the separate ordering of the perforated and whole pebbles, it is possible to recognise whether judges favour the plaintiff or the defendant immediately.

The abacus as a counting tool was used in different modified forms in ancient Greece.[3] The name the Greeks used was probably adopted from the East. The word ἄβαξ (gen. ἄβακος) appears to be of Semitic origin, and refers to sand or dust (קבא, 'ābāq), which suggests that the name derives from a table covered with sand or dust used for drawing or writing,[4] although we lack conclusive evidence.[5] However, it is clear that Latin scholars referenced the Greek tradition and tabular form of the abacus through the name for this counting tool. For example, at the beginning of the 12th century, Tuchill the Computist[6] wrote that this counting table, suitable for calculations in all disciplines of the *quadrivium* – the basis of philosophical knowledge – was formerly (*ab antiquis*) called a *Pythagorean table* (*mensa pytagorica*), whilst in his day (*a modernis*), it was known as an abacus (*abax uel abacus*).[7]

On the cusp between antiquity and the Middle Ages, this counting tool is mentioned by Martianus Capella on several occasions. The abacus (*abacus*) is brought to the wedding feast of Mercury and Philology by Geometry, who, similarly to her sisters (the liberal arts), approaches the celestial wedding gathering

---

2   Ibid. 69, 1, p. 64.
3   See, e.g., Sugden, K. F., "A History of the Abacus." *Accounting Historians Journal* 8/2 (1981), pp. 1–22.
4   Cf. Smith, D. E., *History of Mathematics*. Vol. II: *Special Topics of Elementary Mathematics*. New York: Dover Publications, 1958, pp. 156–157; or Burnett, C. – Ryan, W. F., "Abacus (Western)." In: Bud, R. – Warner, D. J. (eds.), *Instruments of Sciences. An Historical Encyclopedia*. New York – London: Garland Publishing et al., 1998, p. 5.
5   Cf. Pullan, J. M., *The History of the Abacus*. New York: F. A. Praeger Publishers, 1968, p. 17.
6   Cf., for instance, Ambrosetti, N., *L'eredità arabo-islamica nelle scienze e nelle arti del calcolo dell'europa medievale*. Milano: Edizioni Universitarie di Lettere Economia Diritto, 2008, p. 100.
7   Turchill, *Reg. super abacum*, p. 135.

with tools and items typical of her art. This table, covered with dust (*pulvis*),[8] is described as a tool suitable for drawing useful geometrical shapes, e.g., for a description of the universe or the Earth, since it makes it possible to illustrate that which is difficult to express in words.[9] Geometry demonstrates this both at the beginning of her performance for the wedding guests[10] and at the very end[11] when she presents her closing riddle on the abacus (the first sentence of Euclid's *Elements*).[12]

Martianus views the abacus as a table covered with sand that can be used to draw geometric shapes, demonstrating a plenitude of geometrical rules useful for astronomy. Thus, it is not surprising that scholars at early medieval Latin schools often called the abacus *mensa geometricalis*,[13] *tabula geometricalis*,[14] or, possibly referring to the stick used for drawing, *radius geometricus*,[15] since the abacus was mainly used as a tool for the geometrical art.[16] This is clearly evidenced by Remigius of Auxerre, who commented on Martianus' encyclopedia at the end of the 9th century, describing the abacus as a geometrical table (*tabula geometricalis*) covered with greenish (*uitreus*) or bluish (*glaucus*) dust in which geometrical shapes (*figurae geometricae*) are drawn and their appearance is outlined and clearly visible in the dust,[17] which may explain why the name of the tool was derived from dust or sand.

Martianus again mentions the abacus in his compendium when he introduces Arithmetic,[18] a sister of Geometry, to the wedding gathering of celestials. To present her art, she needs a table covered with greenish dust and her oscillating fingers (*digiti recursantes*).[19] Therefore, the abacus is likewise helpful for Arithmetic, whose lecture at the wedding gathering is watched by Pythagoras, seated in the seat of honour.[20] Thus, it is no coincidence that the abacus was commonly

---

8   Martianus, *De nupt.* VI, 587, p. 206.
9   Ibid. VI, 579, pp. 203–204.
10 Ibid. VI, 582, p. 204.
11 Ibid. VI, 724, p. 258.
12 Cf. Euclid, *Elem.* I, prop. 1, pp. 10–12.
13 Remigius, *In Marc.* VI, 288.13, p. 129; *Abacus*, p. 625; or *In Gerb.* I, p. 250 ad.
14 *Reg. arith.*, p. 608; or *Ep. quord.* I, 1, p. 285.
15 Adelard, *Reg. abaci*, p. 91.
16 Cf. also Gerbert's abacus presented by Richer – Richer, *Hist.* III, 54, p. 198.
17 Remigius, *In Marc.* VII, 363.5–6, p. 173; cf. Bernelin, *Liber abaci* I, p. 21.
18 Martianus, *De nupt.* VII, 725, p. 259; or ibid. VII, 729, p. 261.
19 Ibid. VII, 729, p. 261.
20 Ibid. VII, 729, pp. 261–262.

referred to as *mensa Pithagorea* or *mensa Pytagorea*[21] during the Middle Ages. It was Pythagoras who was often mentioned in Latin medieval texts as the first to set down the rules according to which the abacus was used.[22] Besides Pythagoras, the abacus and the development of mathematical art were often ascribed to Archytas of Tarentum[23] or to Plato.[24] Among Latin scholars, Boethius,[25] (and sometimes the Venerable Bede[26]) was often credited with reintroducing this knowledge. The inception of abacus uses in Latin schools by the end of the 10th century is usually tied to Gerbert of Reims,[27] and abacists (*abacisti*) were occasionally referred to as gerbercists (*girbercisti*).[28] An early medieval form of table and column abacus is often referred to as the monastic or the Gerbertian abacus.[29]

Let us now take a brief historical overview of the prominent scholars using the abacus in prescholastic Latin Christian schools. The main initiator of this revival is traditionally believed to have been Gerbert. Around the year 980, Gerbert wrote a brief treatise *Regulae de numerorum abaci rationibus*, which he sent to Constantine at the monastery of Fleury. Not long after this (according to some interpretations, a few years earlier[30]) Constantine's monastic brother in Fleury (and its later abbot) Abbo refers to calculation operations on the abacus

---

21 Adelard, *Reg. abaci*, p. 91; or *Geom. II*, XX, 5, p. 139. It is necessary to mention that this label was also used for another mathematical tool – a multiplication table; for more, see, e.g., Smith, *History of Mathematics…*, pp. 123–128.
22 Odon. *super abacum*, c. 807; Turchill, *Reg. super abacum*, p. 135; or Adelard, *Reg. abaci*, p. 91.
23 See *Geom. II*, XV, 1, p. 135.
24 Ibid. XX, 1, p. 138.
25 *Odon. super abacum*, c. 807B.
26 Ibid., c. 807C.
27 It is evidenced both by some of the oldest descriptions of preserved abacus illustrations in Latin medieval manuscripts (see Figures 27 and 30 below) and by multiple texts from the end of 10th century and 11th century – see, e.g., Bernelin, *Liber abaci*, preaf., p. 16; Ralph, *De abaco*, p. 100; or *Tract. de div.*, p. 630.
28 See, for example, Vatikan, Biblioteca Apostolica Vaticana, MS Vat. Lat. 4539, fol. 74v – URL: https://digi.vatlib.it/view/MSS_Vat.lat.4539 [2019-07-03].
29 See, for instance, Mazur, J., *Enlightening Symbols: A Short History of Mathematical Notation and Its Hidden Powers*. Princeton – Oxford: Princeton University Press, 2014, pp. 48–50; Wedell, M., "Numbers." In: Classen, A. (ed.), *Handbook of Medieval Culture. Fundamental Aspects and Conditions of teh European Middle Ages*. Vol. II. Berlin – Boston: De Gryuter, 2015, pp. 1238–1240; or Folkerts, M. "Frühe Darstellungen des Gerbertschen Abakus." In: Franci – Pagli – Rigatelli (eds.), *Itinera mathematica…*, pp. 23–43, etc.
30 Cf. Bergmann, *Innovationen im Quadrivium…*, p. 180.

in his *Commentary to Calculus*.[31] Most certainly, there were other texts dedicated to abacus counting written in the first millennium, although these appear to be mostly commentaries or additions to Gerbert's *Regulae*.[32] Later, around the year 1000, Gerbert's (probably indirect) pupils, Heriger of Lobbes (*Regulae numerorum super abacum* and *Ratio numerorum abaci*) and Bernelin of Paris (*Liber abaci*) completed longer texts on the abacus and its use.

Apart from several anonymous treatises,[33] abacus literature of the first half of the 11th century was enriched by, for example, a treatise incorporating the first book of the so-called Pseudo-Boethius's *Geometry II*, and by the short texts penned by Hermann of Reichenau (*Qualiter multiplicationes fiant in abaco*) and Laurent of Amalfi (*De divisione*). During the second half of the 11th century and at the beginning of the 12th century, the number of texts on the use of the abacus and the number of abacists increased. Among the most notable were Garland the Computist (*De abaco*), Ralph of Laon (*Liber de abaco*), Adelard of Bath (*Regule abaci*), and Turchill the Computist (*Reguncule super abacum*).

Earlier attempts at a chronological ordering of the oldest abacus texts[34] have been significantly revised in the recent past, thanks to more detailed studies of the surviving depictions of the abacus from the end of the 10th and the 11th century.[35] Today, the prevailing opinion is that it was as early as the end of the 10th century that we can first see the use of Arabic numerals (so-called *ghubar*) in Europe; however, they were only used in abacus calculations. The hypothesis concerning Gerbert's initiating role has once again increased in popularity since Gerbert could have come into contact with *ghubar* numerals during his stay in the shadow of the Pyrenees (see above).

---

31 Abbo, *In Calc.* III, 64–67, pp. 113–115; see also idem, *Excerpta*, pp. 197–203; or idem, *Abacus*, pp. 203–204.
32 See, for example, *De minutiis*, pp. 225–244.
33 See, for instance, *In Gerb.*, pp. 245–284; or the text wrongly attributed to Odo of Cluny – see *Odon. super abacum*, c. 807–814; see also, e.g., other texts edited by B. Boncompagni in *Bulletino di bibliografia e di storia delle scienze matematiche e fisiche* 10 (1877), pp. 595–656, etc.
34 Bergmann, *Innovationen im Quadrivium*..., pp. 205–206.
35 See, e.g., Folkerts, "Frühe Darstellungen...," pp. 23–43; idem, "The names and forms of the numerals on the abacus in the Gerbert tradition." In: Nuvolone, F. G. (ed.), *Gerberto d'Aurillac da Abate di Bobbio a Papa dell'Anno 1000. Atti del Congresso internazionale*. Bobbio: Archivum Bobiense, 2001, pp. 245–265; or Burnett, C., "The Abacus at Echternach in ca. 1000 A.D." *SCIAMVS* 3 (2002), pp. 91–108.

The two oldest surviving Latin descriptions of the abacus will now be presented in detail: Gerbert's abacus as depicted by Richer of Reims, and the *tabula abaci* according to Gerbert's follower Bernelin of Paris; and also several of the oldest schematic drawings representing this calculating tool preserved in Latin manuscripts (originating between the end of the 10th and the beginning of the 12th century).

First, let us explore Richer's description of Gerbert's abacus. In Reims, Gerbert created (with the help of a shield manufacturer) an instrument used to teach geometry. It was a large table divided into 27 columns (*partes*) for the insertion of tokens inscribed with written characters (*notae, caracteres*) of nine digits. Gerbert's abacus had thousands of usable tokens, and, by placing them in the appropriate columns, it was possible to express the value of any number (for values from 10^0 to 10^{26}).[36]

Richer's brief description provides a general idea of what the abacus used by Gerbert looked like. More detailed information is provided in Bernelin's treatise *Liber abaci*, a tract which is explicitly connected to Gerbert, although we know very little about its author.[37] Bernelin conceived almost the entire preface of his treatise as an ode to Gerbert's abacist art[38] (a reference to *Pope Gerbert* is the main argument for dating the text between the years 999 and 1003).[39] He describes the design of the computing tool in detail in the introduction to the first book.

According to *Liber abaci*, the computing tool has 30 columns (*lineae*) with the first three on the right dedicated to the calculation of fractions and the remaining 27 columns used for calculations involving integers. This implies, just as in Richer's description, that the abacus allows the expression of values of numbers from 10^0 to 10^{26}. Each group of three columns is topped by a large arch[40] containing two other arches – a bigger one topping the second and third columns from the right, and a smaller arch topping the first column in each triad.[41]

Each column represents one step in the decimal numerical system. With the help of the arches, this structure was designed to guarantee easier orientation on

---

36   Richer, *Hist.* III, 54, p. 198.
37   See, for example, Bakhouche, B., "Introduction." In: Bernelin, élève de Gerbert d'Aurillac: *Libre d'Abaque...*, p. 9; Bergmann, *Innovationen im Quadrivium...*, p. 199; Lindgren, *Gerbert von Aurillac...*, p. 46.
38   Bernelin, *Liber abaci*, praef., pp. 16–18.
39   Cf., for instance, Folkerts, "The names and forms...," p. 249.
40   Later the arches are called *arcus Pythagorei*.
41   Bernelin, *Liber abaci* I, pp. 21–22.

a large table. For this very reason, the columns are marked by characters – letters explaining the values of the three columns within each large arch. Above the first column on the left, there is the letter 'C' (abbreviation of *centenus* – hundred), indicating that, in this particular position, we are dealing with hundreds (for example, hundreds of millions, hundreds of billions, and so on). The central column is labelled with the letter 'D' (abbreviation of *decenus* – tens), indicating that within the triad of columns, this one represents the values of tens (for example, tens of thousands, tens of billions, and so on). Above the column on the right, there are two letters. The first is 'M' (the abbreviation of the Greek word *monas* – unit), and since it is the mark for the units' column (*unitates*), we have, for example, thousands, quadrillions, and so on. Given the fact that the letter 'M' in descriptions of the individual gradations within an abacus often stands for thousands (*milia*), Bernelin adds an 'S' (abbreviation of *singulares*) after the 'M' to avoid any confusion.[42]

Bernelin then introduces all columns designed for the calculation of integers on the abacist board and transfers their titles to express the individual powers of the decimal notation from 10^0 to 10^{26}. Going from right to left, the units' column is labelled by the letter *I*, the tens column is marked with the letter *X*, and the letter *C* stands for hundreds (from the Roman numerals for one, ten, and hundred); thousands are represented by the letter $\bar{I}$ with a dash (*titulus*) like a 'hat' above the Roman numeral (the values of millions, etc., are all marked by such a 'hat'), $\bar{X}$ with a 'hat' represents tens of thousands, etc., until the twenty-seventh column, which is marked $C\bar{M}\ \bar{M}\bar{M}\ \bar{M}\bar{M}\ \bar{M}\bar{M}\ \bar{M}\bar{I}$ (see Table 13 below).[43]

However, the abacus in Bernelin's description is not only divided vertically into 30 columns, but also horizontally into four horizontal areas (*spatia*), described, from top to bottom, as first, second, third, and fourth.[44] As with the arches or explanatory characters above the individual columns, the horizontal structuring was designed mainly to simplify work with the abacus, allowing various operations in different parts of the table: for example, if we are dividing, there is always a divisor and sometimes a difference in the first area, a dividend in the second area, in the third area calculations and adjustments of the dividend take place, while in the fourth area partial results and the final result (i.e., the quotient) appear. Similarly, in the case of multiplication, the factors are in the first and fourth areas and the product is located in the second area.

---

42   Ibid., pp. 22–23.
43   Ibid., pp. 23–24.
44   Ibid., pp. 24–25.

In compliance with Richer's description of Gerbert's abacus, Bernelin introduces the symbols for numerals (*caracteres, figurae*) used in the abacus. He consistently names the numerals with their Latin designation, followed by Hindu-Arabic (*ghubar*), and Greek numerals (see Table 12 below).[45] Richer's short note and Bernelin's description clearly refer to a very similar design or to the very same device (see Fig. 25). The only difference is that it is unlikely that Gerbert used the abacus for calculations with fractions, due to the absence of the three columns on the right. Richer does not refer to the horizontal division of the abacus nor to the characters marking the columns, although this could be due to the brevity of his account. Heriger of Lobbes wrote his treatises at roughly the same time as Bernelin. However, Heriger does not describe the abacus, but in his computing rules, he incorporates all 27 gradations of the decimal order of magnitude, as shown in Table 13 below.

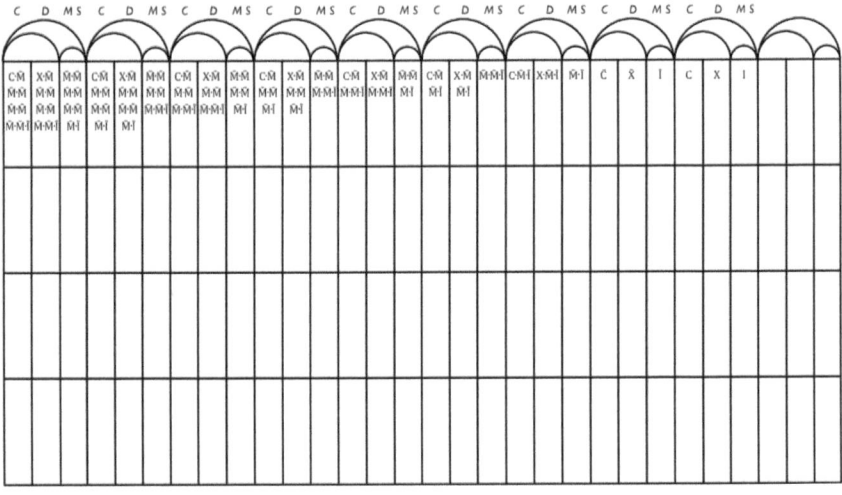

Fig. 25 – Bernelin's abacus according to *Liber abaci* I

Today, there are several surviving period images of the monastic or Gerbertian abacus. The oldest sketches of computing tables date back to the end of the 10th century, and one of the oldest of these sketches is the so-called abacus from Echternach (see Fig. 26), dating back to the late 990s.[46] Images of this abacus have

---

45 Ibid., pp. 25.
46 For description, dating and photocopy see Burnett, "The Abacus at Echternach...," pp. 91–108.

survived in two manuscripts, one as a single sheet inserted into an 11th century Bible (Luxemburg, Bibliothèque nationale de Luxembourg, MS 770), and another, located in the larger manuscript from Echternach, currently displayed in Trier (Trier, Stadtbibliothek, MS 1093/1694, fol. 197r). Both abacus drawings are very similar (the Trier version does not contain the horizontal division of the abacus, and there are slight differences in the depiction of numerals and decimal scales), so we can assume a coincident origin for both drawings. The abacus from the Echternach has 27 columns (the variant on the single sheet has been cut around the edge and the first three right hand columns are missing), each headed with the corresponding decimal scale (the entry in hundreds of thousands is written from millions, i.e., 10^6 to tens of trillions, i.e., 10^{13}). The columns end with arches containing three decimal scales (from units to hundreds, thousands to hundreds of thousands, etc.). Inside these arches, there are smaller arches helping with orientation – always a smaller one for units, thousands, millions, etc., the second, bigger one, tops the remaining two scales within the three-column arch, thus directly corresponding to the description provided by Bernelin. All the smallest arches are marked by the letters *S* (units), and above the abacus, there are the letters *C* (hundreds), *D* (tens), and *M* (units) for each triad of columns (these letters are not present in the Trier variant). Inside the smallest arches, the Western Arabic *ghubar* numerals are written, and the variant from the single sheet also divides the abacus horizontally into four parts.

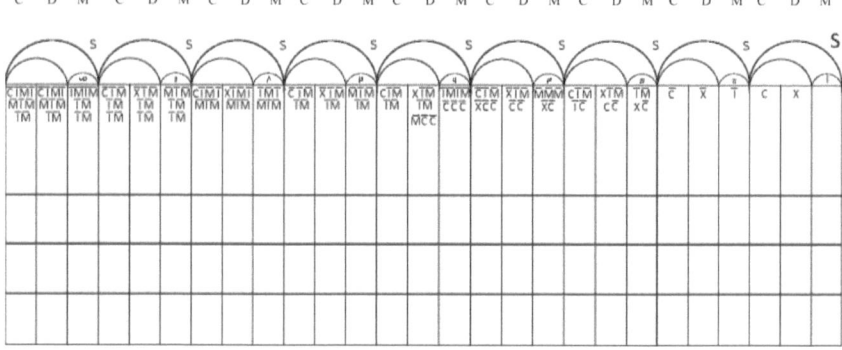

Fig. 26 – Abacus from Echternach (before 1000)
(drawing according to Luxemburg, Bibliothèque nationale de Luxembourg, MS 770 and Trier, Stadtbibliothek, MS 1093/1694, fol. 197r)

During the same period, i.e., at the end of the 10th century, an abacus from a mathematical and computation manuscript from Bern (Bern, Burgerbibliothek,

MS 250, fol. 1r) was drafted. This computing board (see Fig. 27) is perhaps the best-preserved functional abacus from the early Middle Ages. In many ways, it is similar to the abacus from Echtenarch – they share many similar characteristics, such as division into four horizontal parts, large arches topping three columns, with one small arch for units, and a larger one for tens and hundreds, as well as the *C*, *D*, and *M* characters above each triad of arches. However, there are also several differences.[47] The entire computing table is headed with a hexameter providing the information that it was Gerbert who introduced the abacus numerals (i.e., Western Arabic *ghubar* numerals) into the Latin world (*Gerbertus Latio numeros abacique figuras*).[48] However, unlike the abacus from Echtenarch, the verse does not include the shapes of the numerals. The most significant difference is the fact that the Bern abacus consists of 30 rather than 27 columns, allowing calculation of fractions by using the first three columns on the right. These columns are headed with symbols for fractions (*uncia*, *scripulus*, *calcus*), and, naturally, the characters for units, tens, and hundreds (*M*, *D*, and *C*) are not present. Each of the remaining 27 columns is headed with a decimal scale (from 10^0 to 10^{26}) and, similarly to the abacus from Echternach, there is also an alternative with decimal notation in hundreds of thousands. However, in this particular case, the transcript is placed in the second horizontal part of the abacus and proceeds from millions to hundreds of quadrillions (10^{26}). The names, symbols, and mutual ratios of the fractions (from *as* to *calcus*) are stated in the fourth bottom parallel of the abacus. It seems that the author of the drawing had a very in-depth understanding of the skills of abacist computing.

---

47   For description and image see Folkerts, "Frühe Darstellungen…," p. 28 and p. 40, or idem, "The names and forms…," p. 252.
48   See, for example, Folkerts, "Frühe Darstellungen…," p. 28.

Tabula abaci, Decimal Positional Notation and Ghubar Numbers 139

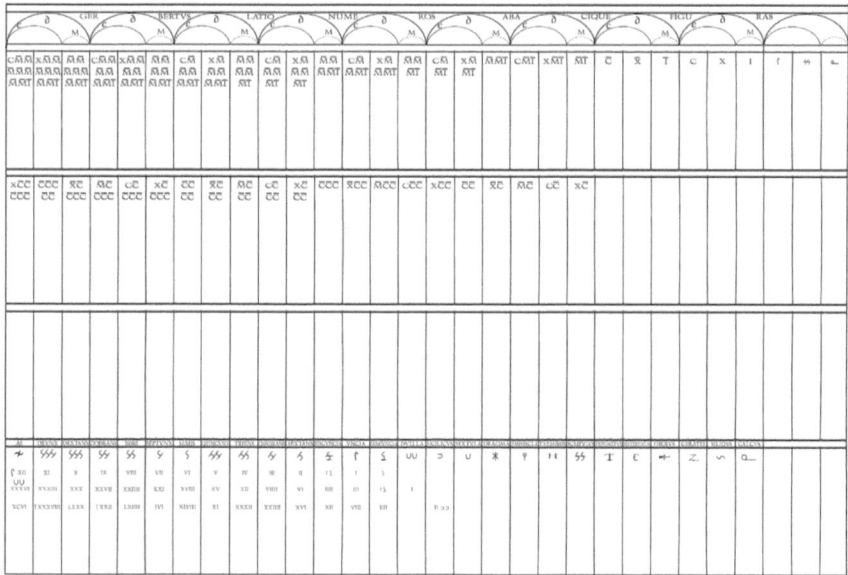

Fig. 27 – The Bern abacus (end of 10th century)
(drawing according to Bern, Burgerbibliothek, MS 250, fol. 1r)

The same cannot be said of the Paris abacus from Fleury, preserved in the astronomy-abacist manuscript from the beginning of 11th century (Paris, Bibliothèque nationale de France, Lat. 8663, fol. 49v).[49] It seems that the author of the abacus drawing originally intended to depict the 27-column abacus (without the columns for fractions), as suggested by the presence of the usual triads of arches (one-, two-, and three-column ones) as well as the fact that they are marked decimally (*C*, *D*, *S* or *M*, with the character *M* perhaps marking thousands as the fourth scale) and titling of the individual scales. After the first three columns (from units to hundreds), the author made a number of mistakes. Instead of 24 columns for the individual decimal scales, he drew only 16, and instead of three columns, only two (see Fig. 28). Therefore, the entire computing instrument becomes very confusing and, in fact, unusable, despite the Arabic numerals written in the smallest arches above the abacus.

---

49 Details and photocopy see in ibid., pp. 28–29, and p. 41 or idem, "The names and forms…," pp. 252–253.

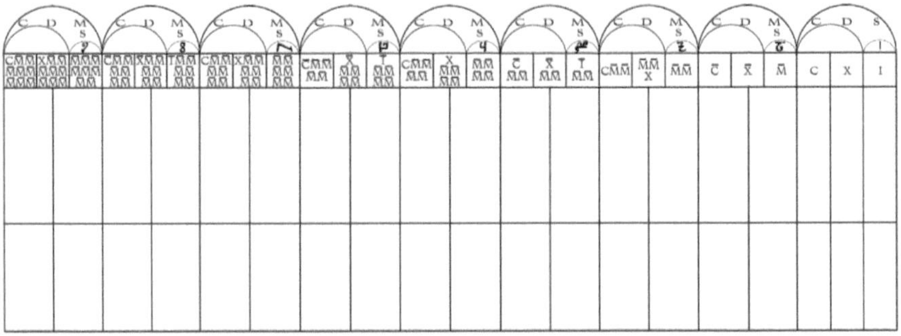

Fig. 28 – The Paris abacus (beginning of 11ᵗʰ century)
(drawing according to Paris, Bibliothèque nationale de France, Lat. 8663, fol. 49v)

The first surviving abacus drawing, which constitutes part of a text treatise on abacus counting, is, at the same time, a pictorial accompaniment to the pseudo-Boethius abacist chapter *Geometria II*, written by an unknown author, most probably from Lorraine (see Fig. 29).[50] Unlike the previous abacuses, this one has only 12 columns (from units to hundreds of billions, i.e., 10^{11}), each topped by its own small arch, ten of which contain the nine Western Arabic numerals and the so-called *sipos*. Above the arches, all ten new symbols are written in words. While the Echternach and Paris abacuses contain nine symbols for numerals (*de facto* directly corresponding to the first nine numerals in the abacus in *Geometria II*), the Pseudo-Boethius's *Geometria II* also contains *sipos*, seemingly reminiscent of our symbol for zero. However, zero was not needed in the abacus: where we would need to write zero today, for example, as in the number 203, the abacist simply left an empty column in the abacus, inserting the symbol for two (*andras*) into the column for hundreds (*C*) and the symbol for the number three (*ormis*) in the units' scale (*I*). *Sipos*, therefore, did not represent zero but an auxiliary symbol, marking the actual place of computing in the abacus table. Furthermore, this abacus is divided into four parts by parallel lines. The symbols of the individual decimal scales are, according to tradition, written in the top part. In general, this drawing serves primarily as an illustration of the text, intended for the purpose of clarification, and, in all

---

50 For details, including photocopies of manuscripts, see Folkerts, M., „Boethius" *Geometrie II. Ein mathematisches Lehrbuch des Mittelalters*. Wiesbaden: Steiner Verlag, 1970, pp. 83–94, and Taf. 1–21.

probability, it does not represent an actual computing instrument to be used for mathematical operations.[51]

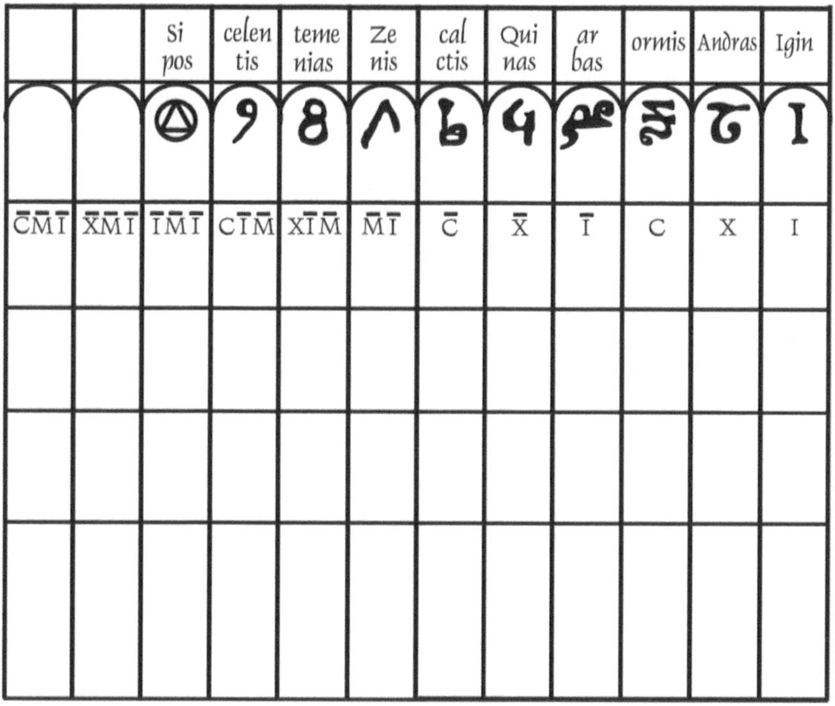

		Si pos	celen tis	teme nias	Ze nis	cal ctis	Qui nas	ar bas	ormis	Anðras	Igin
		⊘	9	8	⋀	Ƅ	4	൚	≊	ნ	1
C̄M̄Ī	X̄M̄Ī	ĪMĪ	CĪM̄	XĪM̄	M̄Ī	C̄	X̄	Ī	C	X	I

Fig. 29 – Abacus from the so-called Pseudo-Boethius's *Geometria II* (first half of 11th century)
(drawing according to Erlangen, Universitätsbibliothek 379, fol. 35r)

The Vatican abacus seems, in many ways, to have been inspired by the Pseudo-Boethius abacus, and constitutes part of a concluding mathematic-abacist insertion in a collection of scientific texts (Vatican, Lat. 644, fols. 77v–78.). Although the manuscript dates back to the 10th century, its last folios with

---

51 A similar case can be found in a mathematical manuscript from the 11th century (Montpellier, H 491, fol. 76r, or fols. 79r–v), including Bernelin's text *Liber abaci* with examples of abacist calculations and Roman fractions. All this is then transferred into accompanying illustrations in the form of abacist tables.

the abacus drawing come from the 11th century.[52] This abacus (see Fig. 30), similarly to the abacus from *Geometria II*, states the names and symbols of the *ghubar* numerals (including the *sipos* assistant mark, which is squeezed into one column together with the number nine, probably to eliminate the possibility of confusing it with zero or ten). Consisting of only 12 columns, it is intended for calculations of up to hundreds of billions (i.e., 10^{11}). It seems that the author of the drawing struggled somewhat (similarly to the author of Paris abacus) since he subsequently added a 13th column for calculations with fractions, as can be seen in the picture. The arches topping the columns are, unusually, written into the individual columns and repeated three times. To a certain extent, this abacus seems the most complex. Besides the columns, which include marks for the individual decimal scales (although the common division with the help of the letters *C, D* and *S/M* is missing, similarly to the abacus from *Geometria II*), and the use of symbols as well as names for Arabic numerals, it also includes the hexameter on Gerbert of Reims and his influence on the use of new numerals among Latin scholars. Furthermore, it contains symbols, names, and mutual ratios of fractions. The hexameter and fractions are also included in the Bern abacus, although in this particular case, the symbols are partially different and, more importantly, the *siliqua* is incorrectly stated (according to Vatican abacus, it represents 1/2 of *calcus*, while usually, it would represent the value of 3/4 of *calcus*, since it is 1/6 of *scripulus*, and *calcus* is 1/8 of *scripulus*). Thus, the value of *sescuncia* is stated incorrectly as well. It is, therefore, possible to assume that the individual who drew this abacus had only partial knowledge of the mathematical processes described.

---

52 Details, including photocopy and references to other editions, in Folkerts, "Frühe Darstellungen…," pp. 29–30, and p. 42; or idem, "The names and forms…," p. 253.

Tabula abaci, Decimal Positional Notation and *Ghubar* Numbers 143

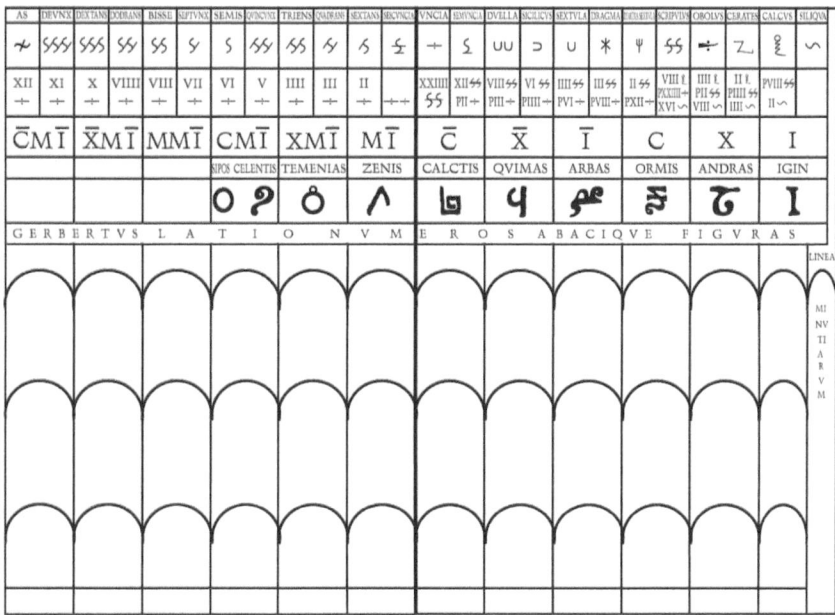

Fig. 30 – The Vatican abacus (11ᵗʰ century)
(drawing according to Vatican, Lat. 644, fols. 77v–78r)

There are several surviving drawings of abacuses from the 11ᵗʰ and 12ᵗʰ centuries.[53] Only three selected cases will be mentioned here. First, there is a manuscript from Fécamp, dating back to the 11ᵗʰ century, and preserved in Rouen (Rouen, Bibliothèque municipale, MS 489, fols. 68v–69r), which contains a drawing of an unusual ten-column abacus (see Fig. 31).[54] Each column ends with an arch with written decimal scale values, symbols, and the names of western-Arabic numerals. Apart from a certain decorativeness, the abacus is particularly interesting due to the presence of Greek numerals (for the values 1–9, i.e., A–Θ). As in the case of the abacuses from Echternach, Bern and *Geometria II*, this abacus is horizontally divided into four parts.

---

53 For example, Oxford, St. John's College, MS 17, fols. 41v–42r. See http://digital.library.mcgill.ca/ms-17/folio.php?p=41v; or http://digital.library.mcgill.ca/ms-17/folio.php?p=42r.
54 Its brief description and photocopy can be found in Folkerts, "The names and forms...," p. 256.

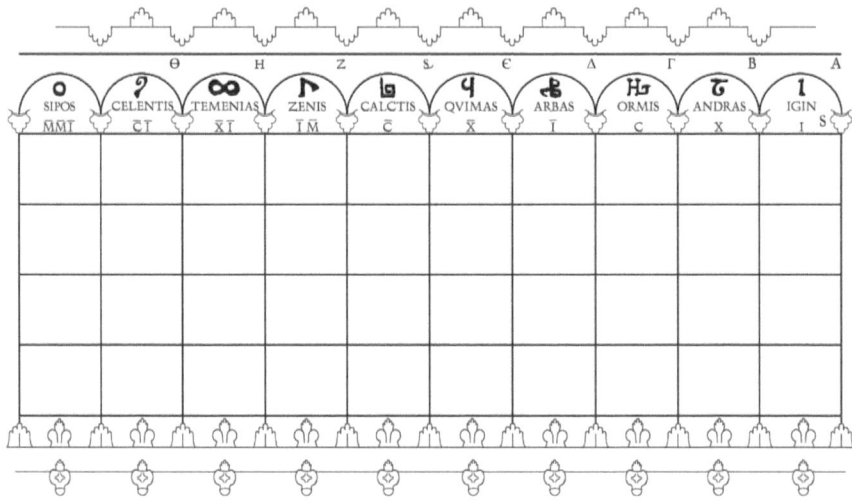

Fig. 31 – The Rouen abacus (11th century)
(drawing according to Rouen, Bibliothèque municipale, MS 489, fols. 68v–69r)

Secondly, there is a manuscript dating back to the 12th century (Paris, Bibliothèque nationale de France, MS Lat. 7231), which summarizes texts dedicated to the liberal arts (mainly rhetoric). In its last folio (85v), there is a drawing of an abacus (see Fig. 32), which dates back to the 11th century (most probably the first half of the century).[55] The reason for such an early dating is its striking similarity to the drawing of the Bern abacus from the late 10th century. Both drawings depict a 30-column abacus with all three sizes of arch topping the columns with the letters *C*, *D*, and *M/S* (the abacus from Bern does not include the *S siglum*), and both are horizontally divided into four sections. In both cases, in the first topmost horizontal section, there are identically written marks for decimal scales from 10^0 to 10^{26} (the second Paris abacus, unlike the Bern abacus, does not include symbols for fractions in the first three columns on the right), and by the boundary line between the first and second horizontal sections, the scales in values of thousands are written in an identical way (in the case of the second Paris abacus, they are written above the borderline, i.e., in the bottom part of the first section, while in the case of the Bern abacus, they are written under the dividing line, i.e., in the upper part of the second section).

---

55   See, e.g., Burnett, "The Abacus at Echternach…," p. 92.

In both cases, the fourth section represents fractions, their names (the author of the second Paris abacus distorted the name of *sestertius*), symbols (there are only minute differences in both drawings), and mutual ratios between fractions. However, there are several significant differences between both drawings. While the Bern abacus states the triple numeral values of *as* (*uncia, duella* and, on two occasions, *sicilicus*), the second Paris abacus focuses on *scripulus* and its relationship to *as* or, in the case of smaller values, it adds mutual values *scripulus–obolus* and *obolus–cerates–calcus*. While the second Paris abacus does not include the hexameter on Gerbert's introduction of the *ghubar* numerals to the Latin West, it allows for values determined by the number five to be inserted into the relevant columns, i.e., the semi-values of the decimal scales by the borderline between the second and third horizontal section. Here, the scribe made an error since, in the third column on the left (10^{24}), he wrote the same characters as for the decimal values instead of the half-value (correctly, it should be V̄C̄C̄C̄C̄). Furthermore, near the dividing line between the first and second horizontal sections, ahead of the markings of the columns with values of hundreds of thousands, the second Paris abacus expresses the values of the columns with the help of the Roman numeral for five (scales 10^1 to 10^5) and the scale of units (10^0) as twice the value of the *semis*. Another error in the second Paris abacus is the fact that the symbols *C, D,* and *M* are written in the first three columns on the right dedicated to calculations of fractions and, therefore, devoid of the need for written decimal values. Finally, in the smallest arches, the Western Arabic numerals are included in the second Paris abacus and omitted from the Bern abacus. It seems, therefore, that the drawing of the second Paris abacus was strongly inspired by the Bern abacus; however, it is not simply a copy of the Bern abacus, as the author of the second Paris abacus added some important and interesting amendments, albeit not always correctly.

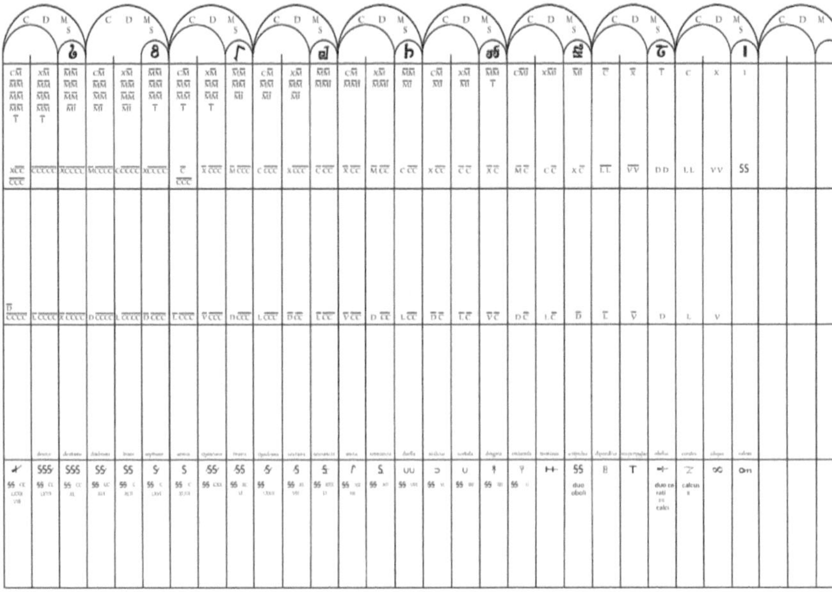

Fig. 32 – The second Paris abacus (11ᵗʰ century)
(drawing according to Paris, Bibliothèque nationale de France, Lat. 7231, fol. 85v)

The final manuscript analysed here is a computistic manuscript from St. John's College, Oxford, dating back to the first quarter of the 12ᵗʰ century. It contains a particularly decorative drawing of a 27-column abacus (Oxford, St. John's College, MS 17, fols. 48v–49r).[56] Of all the abacuses discussed in this chapter, this is the 'newest', and, at the same time, it represents a very precisely executed computing tool. Besides the columns for calculations with fractions, it also communicates all other instructional information (see Fig. 33): every column is marked with the value of the decimal scale; a large arch always tops the triad of columns (with the characters C, D and M/S as in Echternach, Bern

---

56 The abacus is available online: http://digital.library.mcgill.ca/ms-17/folio.php?p=48v; or http://digital.library.mcgill.ca/ms-17/folio.php?p=49r. For its analysis, and interpretation, including photocopies, see Evans, G. R., "Difficillima et Ardua: theory and practice in treatises on the abacus, 950–1150." *Journal of Medieval History* 3 (1977), pp. 21–38; or eadem, "Schools and scholars: the study of the abacus in English Schools c. 980–c. 1150." *The English Historical Review* 94 (1979), pp. 71–89.

and both Paris abacuses) with two other smaller arches inside – one larger than the other, topping two columns, and the second smaller arch for units (just as in Bernelin's description); in the smallest arch, the Western Arabic, Greek, and Roman numerals are inscribed with the names of Hindu-Arabic numerals above. Although the columns dedicated to calculations with fractions do not display Roman fractions, their symbols, names, and some of their ratios are written in the bottom section of the abacus (and in a departure from the Vatican abacus, the *siliqua* is stated here correctly, as well as its multiples, the *bissiliqua* and the *tremissis*).

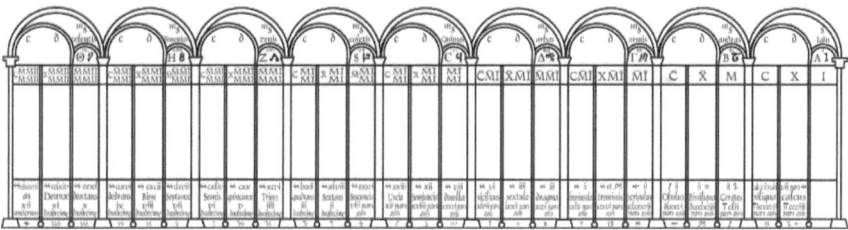

Fig. 33 – The Oxford abacus (around the year 1110)
(drawing according to Oxford, St. John's College, MS 17, fols. 48v–49r)

Although some of the abacus drawings discussed above contain incorrect information or present a computing table with a confused or fundamentally unworkable structure, it is evident that the majority of the examples correspond with Richer's and Bernelin's descriptions of the tool. Moreover, all presented diagrams of the abacus attempt to make computing practice as easily understandable as possible for the users and readers alike. However, it is clear that the authors of these diagrams did not always properly grasp all aspects of abacist calculations themselves, especially the Hindu-Arabic symbols for numerals (or their Roman and Greek names and equivalents); the efforts put into the orientation markings of the decimal notation, and into the symbols and mutual ratios of fractions, were intended to make the tool easier to use.

All in all, having examined the various depictions of abacuses, it is possible to finally summarize the basic information conveyed by the oldest diagrams of abacuses, drawn in the oldest early Medieval Latin texts dedicated to abacist mathematics. First of all, these drawings constitute the first introduction of Western Arabic numerals into the Western Christian world. The symbols for these numerals varied greatly, as can be observed primarily in the case of the numeral '3', as well as in the different positions of some symbols, due to the fact that the tokens with the numeral symbol could be inserted into the columns in any position, as it did not represent a major complication for the abacist (see Table 12).

Table 12 – Symbols and names of *ghubar*-numerals

numerals	Greek	Roman	abacist symbols	names
1	A	I	1 1	igin
2	B	II	ɼ	andras
3	Γ	III	₹ ₹ Hɟ ₴ m̃	ormis
4	Δ	IIII (IV)	℘ ⅃ ๙ ⁏	arbas
5	E	V	५ ҍ	quinas/quimas
6	Σ	VI	ᗡ Ⴆ ៤ ₫	calctis
7	Z	VII	⋀ ⋂ ⋔	zenis
8	H	VIII	8 ȯ ∞	temenias
9	Θ	VIIII (IX)	⅏ ʔ 9 ƃ	celentis
	auxiliary mark		O ⊙	sipos

In the past, serious doubts concerning Gerbert's own use of such *ghubar* numerals on the abacus were voiced.[57] However, such reservations were not met with widespread acceptance for a number of reasons.[58] One proof of Gerbert's knowledge of Hindo-Arabic numerals is represented by a panegyric figurative poem (see Fig. 34) which was apparently created in the middle of the year 983 (when Gerbert was the abbot of the monastery in Bobbio) on the occasion of the coronation of (the nearly three-year-old) Otto III. The poem was identified as Gerbert's work at the end of the last century, when the individual verses were ordered into the form of a grid poem (see Fig. 35), into which an acrostic is imbedded, ending with Gerbert's dedication to the imperial husband and wife Otto II and Theophano (*a Gerberto Ottoni ac Theophano*).[59]

---

57   See Bergmann, *Innovationen im Quadrivum…*, pp. 194–195.
58   Cf., for example, Folkerts, "The names and forms…," pp. 246–247; or Burnett, "The Abacus at Echternach…," pp. 92–93; etc.
59   See, especially, Brockett, W., "The Frontispiece of Paris, Bibliothèque Nationale Ms. Lat. 776. Gerbert's Acrostic Pattern Poems." *Manuscripta* 39 (1995), pp. 3–25; Nuvolone,

Fig. 34 – Gerbert's figurative panegyric poem for Otto II

F. G., "Il *Carmen* figurato attribuito a Gerberto nel Ms Paris, BNF, lat. 776, fol. 1v: una composizione redatta nell'Abbazia di San Colombano di Bobbio?." *Archivum Bobiense* 24 (2002), pp. 123–260; or idem, "Quelques éléments d'introduction au *Carmen Figuratum* de Gerbert d'Aurillac pubblicato." In: Sigismondi, C. (ed.), *Culmina Romulea: fede e scienza in Gerberto, papa filosofo*. Roma: Ateneo Pontificio Regina Apostolorum, 2008, pp. 47–83. The poem is also edited as Gerbert, *Figur.* X, pp. 666–667.

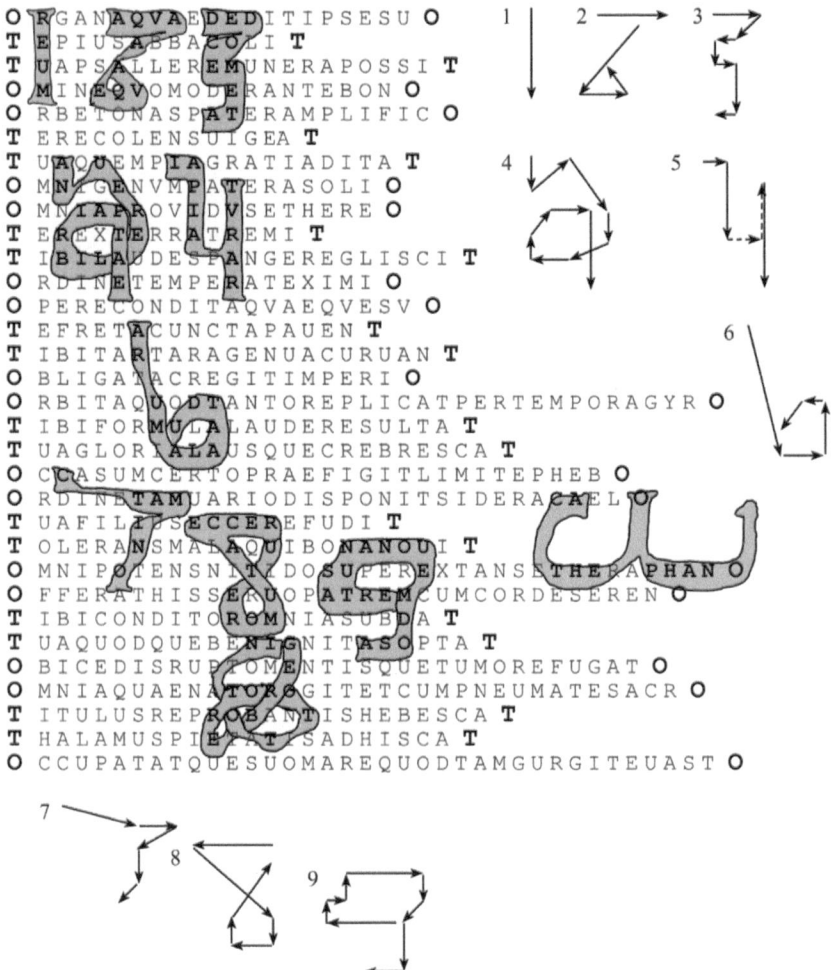

```
O RGANAQVAEDEDITIPSESU O        1 ↓   2 →   3 →
T EPIUSABBACOLI T
T UAPSALLEREMUNERAPOSSI T
O MINEQVOMODERANTEBON O
O RBETONASPATERAMPLIFIC O
T ERECOLENSUIGEA T
T UAQUEMPIAGRATIADITA T          4 ↓         5 →
O MNIGENVMPATERASOLI O
O MNIAPROVIDVSETHERE O
T EREXTERRATREMI T
T IBILAUDESPANGEREGLISCI T
O RDINETEMPERATEXIMI O
O PERECONDITAQVAEQVESV O
T EFRETACUNCTAPAUEN T                            6
T IBITARTARAGENUACURUAN T
O BLIGATACREGITIMPERI O
O RBITAQUODTANTOREPLICATPERTEMPORAGYR O
T IBIFORMULALAUDERESULTA T
T UAGLORIALAUSQUECREBRESCA T
O CEASUMCERTOPRAEFIGITLIMITEPHEB O
O RDINETAMUARIODISPONITSIDERACAEL O
T UAFILIISECCEREFUDI T
T OLERANSMALAQUIBONANOUI T
O MNIPOTENSNITIDOSUPEREXTANSETHERAPHAN O
O FFERATHISSERUOPATREMCUMCORDESEREN O
T IBICONDITOROMNIASUBDA T
T UAQUODQUEBENIGNITASOPTA T
O BICEDISRUPTOMENTISQUETUMOREFUGAT O
O MNIAQUAENATOROGITETCUMPNEUMATESACR O
T ITULUSREPROBANTISHEBESCA T
T HALAMUSPIETATISADHISCA T
O CCUPATATQUESUOMAREQUODTAMGURGITEUAST O
```

Fig. 35 – Gerbert's figurative panegyric poem for Otto II as a grid poem with acrostic

The acrostic verses of the grid poem represent the shapes of Western Arabic numerals. Should readers follow the ordering of letters for individual numbers, they will also receive instructions for writing them. The very shape of these numerals closely corresponds both to the oldest preserved form of Western Arabic numerals in the Latin text from Vigila's manuscript[60] and to abacist symbols for numbers from the early medieval pictures of abacuses. It might be surmised that Gerbert intentionally composed the poem to show off one of the innovations he had introduced in Latin schools. To the nine numerical figures, which could have been the *figurae* mentioned by Richer of Reims in relation to Gerbert's abacus, a *triquetra* and Greek omega, i.e., Greek numeral 800, are added, possibly to symbolise the restoration of the Western Empire by the coronation of Charlemagne. Thus, it is undoubtedly not a coincidence that the poem consists of precisely 800 letters and its verses compellingly capture the Ottonian vision of the order of contemporary Europe and the role and stature of the emperor, including the legitimacy of his power derived from God Himself.[61] If this interpretation is correct, it is clear that Gerbert knew Arabic numerals, and he used tokens with these or similar shapes for numerals in abacus counting.

Another difficult task for the abacist was to gain a better grasp of the large table and become familiar with making entries of the numeral values according to the decimal notation. This task would have been facilitated by the headings showing the Roman numerals of decimal scales above the individual columns. The symbols for the higher decimal scales varied considerably (see Table 13) and would often be used incorrectly. However, the principle of marking was, in fact, identical. Some of the abacus drawings showed a decimal scale value in the columns not only in thousands, millions, etc., but also in hundreds of thousands.

---

60   The monk Vigila lived during the last third of the 10th century in the north of the Iberian Peninsula in the monastery of San Martín de Albelda. In his manuscript (known as *Codex Vigilanus sive Albeldensis*, i.e., El Escorial, Real Biblioteca del Monasterio de San Lorenzo, Ms d.I.2.) from the year 976, we can find the very first use of *ghubar* numerals in Latin text. They appear in a commentary to the fourth chapter of the third book of Isiodor's *Etymologiae* about the usefulness of numbers.
61   For more details, see, e.g., Nuvolone, F. G., "Numeri, Croce e Vita: Gerberto e la Parola. A proposito della rilegatura di Echternach: un programma Gerbertiano?," *GERBERTVS – International Academic Publication on History of Medieval Science* 1 (2010), pp. 110–169.

Table 13 – Markings of the abacus columns

scale	English	Bernelin	Echternach abacus	Bern abacus	Paris abacus	Oxford abacus
10^0	one	I (S, M)	I (S)	I (M)	I (S)	I (S)
10^1	ten	X (D)	X	X (D)	X (D)	X (D)
10^2	hundred	C	C	C	C	C
10^3	thousand	$\bar{\text{I}}$ (M)	$\bar{\text{I}}$	$\bar{\text{I}}$	$\bar{\text{M}}$	M
10^4	ten thousand	$\bar{\text{X}}$	$\bar{\text{X}}$	$\bar{\text{X}}$	$\bar{\text{X}}$	$\bar{\text{X}}$
10^5	hundred thousand	$\bar{\text{C}}$	$\bar{\text{C}}$	$\bar{\text{C}}$	$\bar{\text{C}}$	$\bar{\text{C}}$
10^6	million	$\bar{\text{M}}\bar{\text{I}}$	$\bar{\text{I}}\bar{\text{M}}$ / X $\bar{\text{C}}$	$\bar{\text{M}}\bar{\text{I}}$ / X $\bar{\text{C}}$	$\bar{\text{M}}\bar{\text{M}}$	$\bar{\text{M}}\bar{\text{I}}$
10^7	ten million	X $\bar{\text{M}}\bar{\text{I}}$	X $\bar{\text{I}}\bar{\text{M}}$ / C $\bar{\text{C}}$	X $\bar{\text{M}}\bar{\text{I}}$ / C $\bar{\text{C}}$	X $\bar{\text{M}}\bar{\text{M}}$	X $\bar{\text{M}}\bar{\text{I}}$
10^8	hundred million	C $\bar{\text{M}}\bar{\text{I}}$	C $\bar{\text{I}}\bar{\text{M}}$ / $\bar{\text{I}}$ $\bar{\text{C}}$	C $\bar{\text{M}}\bar{\text{I}}$ / $\bar{\text{M}}$ $\bar{\text{C}}$	C $\bar{\text{M}}\bar{\text{M}}$	C $\bar{\text{M}}\bar{\text{I}}$
10^9	billion	$\bar{\text{M}}\bar{\text{M}}\bar{\text{I}}$	$\bar{\text{M}}$ $\bar{\text{M}}\bar{\text{M}}$ / $\bar{\text{X}}$ $\bar{\text{C}}$	$\bar{\text{M}}\bar{\text{M}}\bar{\text{I}}$ / $\bar{\text{X}}$ $\bar{\text{C}}$	$\bar{\text{I}}$ $\bar{\text{M}}$ M	$\bar{\text{M}}\bar{\text{M}}\bar{\text{I}}$
10^{10}	ten billion	X $\bar{\text{M}}\bar{\text{M}}\bar{\text{I}}$	$\bar{\text{X}}$ $\bar{\text{I}}\bar{\text{M}}$ / $\bar{\bar{\text{C}}\text{C}}$	X $\bar{\text{M}}\bar{\text{M}}\bar{\text{I}}$ / $\bar{\bar{\text{C}}\text{C}}$	$\bar{\text{X}}$ $\bar{\text{M}}\bar{\text{M}}$	$\bar{\text{X}}\bar{\text{M}}\bar{\text{I}}$

Table 13 Continued

scale	English	Bernelin	Echternach abacus	Bern abacus	Paris abacus	Oxford abacus
10^{11}	hundred billion	C M̄M̄Ī	C̄ ĪM X C̄C̄	C M̄M̄Ī X C̄C̄	C̄ M̄M̄	C̄M̄Ī
10^{12}	trillion	M̄M̄M̄Ī	ĪM ĪM C̄C̄	M̄M̄M̄Ī C C̄C̄	M̄M̄ M̄M̄	M̄Ī M̄Ī
10^{13}	ten trillion	X M̄M̄M̄Ī	X ĪM ĪM M̄C̄C̄	X M̄M̄M̄Ī M̄ C̄C̄	X M̄M̄ M̄M̄	X M̄Ī M̄Ī
10^{14}	hundred trillion	C M̄M̄M̄Ī	C ĪM ĪM	C M̄M̄M̄Ī X̄ C̄C̄	C M̄M̄ M̄M̄	C M̄Ī M̄Ī
10^{15}	quadrillion	M̄M̄M̄M̄Ī	M̄ ĪM ĪM	M̄M̄M̄M̄Ī C̄ C̄C̄	Ī M̄M̄ M̄M̄	M̄ M̄Ī M̄Ī
10^{16}	ten quadrillion	X M̄M̄M̄M̄Ī	X̄ ĪM ĪM	X M̄M̄M̄M̄Ī X C̄ C̄C̄	X̄ M̄M̄ M̄M̄	X̄ M̄Ī M̄Ī
10^{17}	hundred quadrillion	C M̄M̄M̄M̄Ī	C̄ ĪM ĪM	C M̄M̄M̄M̄Ī C C̄C̄	C̄ M̄M̄ M̄M̄	C̄ M̄Ī M̄Ī
10^{18}	quintillion	M̄M̄M̄M̄M̄Ī	ĪM ĪM ĪM	M̄M̄M̄M̄M̄Ī M̄ C̄C̄C̄	M̄M̄ M̄M̄ M̄M̄	M̄M̄Ī M̄M̄Ī
10^{19}	ten quintillion	X M̄M̄M̄M̄M̄Ī	X ĪM ĪM ĪM	X M̄M̄M̄M̄M̄Ī X̄ C̄C̄C̄	X M̄M̄ M̄M̄ M̄M̄	X M̄M̄Ī M̄M̄Ī

(continued on next page)

154 Practical Arithmetic

scale	English	Bernelin	Echternach abacus	Bern abacus	Paris abacus	Oxford abacus
10^{20}	hundred quintillion	$C\bar{M}\bar{M}\bar{M}\bar{M}\bar{M}\bar{M}\bar{I}$	$C\ \bar{I}\bar{M}\ \bar{I}\bar{M}\ \bar{I}\bar{M}$	$C\bar{\bar{M}}\bar{\bar{M}}\bar{\bar{M}}\bar{\bar{M}}\bar{M}\bar{I}$ $\bar{\bar{C}}\bar{\bar{C}}$	$C\ \bar{M}\bar{M}\ \bar{M}\bar{M}\ \bar{M}\bar{M}$	$C\ \bar{M}\bar{M}\bar{I}\ \bar{M}\bar{M}\bar{I}$
10^{21}	sextillion	$\bar{M}\bar{M}\bar{M}\bar{M}\bar{M}\bar{M}\bar{I}$	$\bar{M}\ \bar{I}\bar{M}\ \bar{I}\bar{M}\ \bar{I}\bar{M}$	$\bar{M}\bar{\bar{M}}\bar{\bar{M}}\bar{\bar{M}}\bar{\bar{M}}\bar{M}\bar{I}$ $\bar{\bar{X}}\ \bar{\bar{C}}\bar{\bar{C}}$	$\bar{I}\ \bar{M}\bar{M}\ \bar{M}\bar{M}\ \bar{M}\bar{M}$	$\bar{M}\ \bar{M}\bar{M}\bar{I}\ \bar{M}\bar{M}\bar{I}$
10^{22}	ten sextillion	$X\bar{M}\bar{M}\bar{M}\bar{M}\bar{M}\bar{M}\bar{I}$	$\bar{X}\ \bar{I}\bar{M}\ \bar{I}\bar{M}\ \bar{I}\bar{M}$	$X\bar{\bar{M}}\bar{\bar{M}}\bar{\bar{M}}\bar{\bar{M}}\bar{M}\bar{I}$ $C\ \bar{\bar{C}}\bar{\bar{C}}$	$\bar{X}\ \bar{M}\bar{M}\ \bar{M}\bar{M}\ \bar{M}\bar{M}$	$\bar{X}\ \bar{M}\bar{M}\bar{I}\ \bar{M}\bar{M}\bar{I}$
10^{23}	hundred sextillion	$C\bar{M}\bar{M}\bar{M}\bar{M}\bar{M}\bar{M}\bar{I}$	$\bar{C}\ \bar{I}\bar{M}\ \bar{I}\bar{M}\ \bar{I}\bar{M}$	$C\bar{\bar{M}}\bar{\bar{M}}\bar{\bar{M}}\bar{\bar{M}}\bar{M}\bar{I}$ $\bar{M}\ \bar{\bar{C}}\bar{\bar{C}}$	$\bar{C}\ \bar{M}\bar{M}\ \bar{M}\bar{M}\ \bar{M}\bar{M}$	$\bar{C}\ \bar{M}\bar{M}\bar{I}\ \bar{M}\bar{M}\bar{I}$
10^{24}	septillion	$\bar{M}\bar{M}\bar{M}\bar{M}\bar{M}\bar{M}\bar{I}$	$\bar{I}\bar{M}\ \bar{I}\bar{M}\ \bar{I}\bar{M}\ \bar{I}\bar{M}$	$\bar{M}\bar{\bar{M}}\bar{\bar{M}}\bar{\bar{M}}\bar{\bar{M}}\bar{M}\bar{I}$ $\bar{\bar{X}}\ \bar{\bar{C}}\bar{\bar{C}}$	$\bar{M}\bar{M}\ \bar{M}\bar{M}\ \bar{M}\bar{M}\ \bar{M}\bar{M}$	$\bar{M}\bar{M}\ \bar{I}\bar{I}\ \bar{M}\bar{M}\ \bar{I}\bar{I}$
10^{25}	ten septillion	$X\bar{M}\bar{M}\bar{M}\bar{M}\bar{M}\bar{M}\bar{I}$	$X\ \bar{I}\bar{M}\ \bar{I}\bar{M}\ \bar{I}\bar{M}\ \bar{I}\bar{M}$	$X\bar{M}\bar{\bar{M}}\bar{\bar{M}}\bar{\bar{M}}\bar{\bar{M}}\bar{M}\bar{I}$ $\bar{\bar{C}}\ \bar{\bar{C}}\bar{\bar{C}}$	$X\ \bar{M}\bar{M}\ \bar{M}\bar{M}\ \bar{M}\bar{M}\ \bar{M}\bar{M}$	$X\ \bar{M}\bar{M}\ \bar{I}\bar{I}\ \bar{M}\bar{M}\ \bar{I}\bar{I}$
10^{26}	hundred septillion	$C\bar{M}\bar{M}\bar{M}\bar{M}\bar{M}\bar{M}\bar{I}$	$C\ \bar{I}\bar{M}\ \bar{I}\bar{M}\ \bar{I}\bar{M}\ \bar{I}\bar{M}$	$C\bar{M}\bar{\bar{M}}\bar{\bar{M}}\bar{\bar{M}}\bar{\bar{M}}\bar{M}\bar{I}$ $X\bar{\bar{C}}\ \bar{\bar{C}}\bar{\bar{C}}$	$C\ \bar{M}\bar{M}\ \bar{M}\bar{M}\ \bar{M}\bar{M}\ \bar{M}\bar{M}$	$C\ \bar{M}\bar{M}\ \bar{I}\bar{I}\ \bar{M}\bar{M}\ \bar{I}\bar{I}$

The final common auxiliary tool of the abacist was a set of names, symbols, and mutual ratios between fractions. Arithmetical operations with fractions were, along with division, regarded as a high-level skill, which could be mastered only by an experienced mathematician. The lack of this skill would be partially rectified with the help of written overviews placed in the bottom part of the board (see Table 14).

Table 14 – Fractions according to the drawings of early medieval abacuses

	as	uncia	scripulus	calcus
as	1	12	288	2304
deunx	11/12	11	264	2112
dextans	5/6	10	240	1920
dodrans	3/4	9	216	1728
bisse	2/3	8	192	1536
septunx	7/12	7	168	1344
semis	1/2	6	144	1152
quinqunx	5/12	5	120	960
triens	1/3	4	96	768
quadrans	1/4	3	72	576
sextans	1/6	2	48	384
sexcuncia	1/8	3/2	36	288
uncia	1/12	1	24	192
semiuncia	1/24	1/2	12	96
duella	1/36	1/3	8	64
sicilicus	1/48	1/4	6	48
dragma (sextula)	1/72	1/6	4	32
(h)emisescla (dimidia sextula)	1/144	1/12	2	16
tremissis	1/216	1/18	3/2	12
scripulus	1/288	1/24	1	8
obolus	1/576	1/48	1/2	4
bissiliqua	1/864	1/72	1/3	8/3
cerates	1/1152	1/96	1/4	2
siliqua	1/1728	1/144	1/6	4/3
calcus	1/2304	1/192	1/8	1

Without any doubt, abacist counting represented a significant change in fundamental arithmetical operations. It was not easy for intellectuals of that period to become familiar with the subject of abacism. Therefore, the oldest Latin texts

2. *Regulae multiplicationis*

During the second half of the 990s, Emperor Otto III received a letter from Gerbert wishing him to live to an age corresponding to the highest number that could be expressed on an abacus (*extremus numerorum abbaci*).[62] The number meant by the author of this remarkable message is 999 999 999 999 999 999 999 999, which is the highest value that the early medieval abacus with 27 columns (i.e., orders of integers) can express. If the abacist table used by scholars around the year 1000 had been more extensive, the highest possible number would have been greater. In order to understand Gerbert's statement, the Emperor would have had to be well informed about the method of expressing quantitative values on an abacus, which might well relate to the fact that Gerbert stayed at the imperial court and is rightfully considered to be the initiator of a growing interest in the abacus at early medieval Latin schools.

Apart from Richer's aforementioned description of Gerbert's abacus, two versions of a brief account of the counting rules for an abacus have been preserved, the so-called *Regulae de numerorum abaci rationibus*. Gerbert wrote the first briefer version at the end of the 970s and attached it to the letter to Constantine.[63] In this letter, he wrote that he had lacked access to books on the issue for many years (*aliquot lustra*), but he had attempted to write down what he could remember from before.[64] This might imply that he had learned to work with the abacus or abacist treatises and, in particular, with Arabic numerals and the Arabic way of recording numerical values during his earlier Catalonian

---

62 Gerbert, *Ep.* 183, p. 217.
63 It concerns the *Textus Gerberti genuinus*, as titled by the editor of Gerbert's treatise N. Bubnov. For a dating respected to this day, see Bergmann, *Innovationen im Quadrivum*..., p. 180. The second expanded version of the text, known as *Textus interpolates*, might not be Gerbert's work, according to Bubnov, but it may have originated in his circle or in response to the need to clarify and flesh out some disputed passages from the first version of the work. It was supposedly written around the year 1000 or, more likely, shortly after – cf., e.g., Folkerts, M., "The *Geometry* II Ascribed to Boethius." In: idem, *Essays on Early Medieval Mathematics. The Latin Tradition*. Aldershot: Ashgate, 2003, p. 2, and p. 8.
64 Gerbert, *Regulae*, praef., p. 6.

studies.⁶⁵ This hypothesis is likewise supported by Gerbert's letter to Gerald the abbot of the Aurillac abbey from the year 984, in which Gerbert asks the abbot for a book by the otherwise unknown Joseph of Spain (*Ioseph Ispano*) *De multiplicatione et divisione numerorum*, which had supposedly been left in the Aurillac monastery by his friend from his Catalonian studies Garin (*Warnerius, Guarnerius*), the abbot of the abbey of Saint-Michel-de-Cuxa.⁶⁶ In the same year, Gerbert wrote to another Catalonian friend Miró Bonfill, Count of Besalú and bishop of Girona, that he would like to obtain the book of the same title, written by Joseph the Wise (*Ioseph sapiens*) for the Reims library.⁶⁷ It seems that in both letters, Gerbert requested the very same book about multiplication and division. This manuscript (unknown to us today) may have been the primary source for Gerbert when he studied counting operations on the abacus, and when he later formulated his own rules for these operations.⁶⁸

The work of Joseph of Spain/Joseph the Wise might not have been the only treatise on abacist counting known to Gerbert. The aforementioned *Codex Vigilanus* and its folio 12v, in which the forms of Arabic numerals are listed and praised, presented rules of multiplication which are strikingly similar to Gerbert's multiplication rules from his *Regulae*.⁶⁹ These facts indirectly support the thesis that Gerbert was one of the initiators of a growing interest in modified abacist computing, especially in Europe beyond the Pyrenees during the last third of the 10th century, and that he was one of the first promoters of Arabic *ghubar* numerals, which he used for abacus computations.⁷⁰

While Gerbert's text concerning computing on an abacus is relatively brief, other contemporary texts dealing with the same topic (and often conceived as commentaries or addendums to Gerbert's treatise – e.g., *Regulae numerorum*

---

65 Cf., for instance, Silva, J. N., "O Ábaco de Gerbert." *GERBERTVS – International Academic Publication on History of Medieval Science* 4 (2013), pp. 102–103; or Schärling, A., *Un portrait de Gerbert d'Aurillac…*, pp. 54–60.
66 Gerbert, *Ep.* 17, p. 40.
67 Gerbert, *Ep.* 25, p. 48.
68 Cf, for example, Saenger, P., *Space between Words. The Origins of Silent Reading.* Stanford: Stanford University Press, 1997, p. 130.
69 The passage dealing with multiplication was edited by H. Pratt Lattin – see Lattin, Pratt, H., "The origin of our present system of notation according to the theories of Nicholas Bubnov." *Isis* 19/1 (1933), pp. 191–192. Cf. also, e.g., Folkerts, "The names and forms…," p. 247; or idem, "Early Texts on Hindu-Arabic Calculation." *Science in Context* 14/1–2 (2001), p. 16.
70 Cf. Catalani, "Omnia Numerorum…," p. 138.

super abacum and *Ratio numerorum abaci* by Heriger of Lobbes or *Liber abaci* by Bernelin of Paris; while the abacist activities of Abbo of Fleury also deserve mention[71]) are generally more thorough and it is possible to identify in them a standardised form:

1. a description of the abacus and the introduction of the basic terminology employed when using an abacus;
2. rules of multiplication;
3. rules of division;
4. an analysis of computing operations with fractions.

Of all the works originating around the year 1000, the only completely preserved description of all four parts is that of Bernelin. The accounts of Gerbert and Heriger lack the descriptive illustration of the abacus and instructions on computing with fractions, whereas that of Abbo of Fleury does not include, for example, the rules of division.

Given that the abacist computations were not meant only for scientific or mathematical purposes (early medieval arithmeticians could have conducted most of the computing operations more easily by finger-counting),[72] but were also meant to entertain those making the computations,[73] we will further focus on the counting operations themselves. First, we will deal with multiplication and, in the next subchapter, we will shift our attention to the delight of every mathematics scholar – i.e., division.

---

71 Heriger of Lobbes and Bernelin of Paris can apparently be considered as Gerbert's direct or indirect disciples, who referenced their teacher in their treatises. Abbo of Fleury knew Gerbert personally; however, in many respects, they were rivals who assumed opposing positions. Nevertheless, it is known to us that Abbo was called *abaci doctor* (cf. Peden, "Introduction," p. xxxviii; or Abbo, *Excerpta*, p. 203), and it is beyond doubt that he understood counting operations on an abacus (*In Calculum Victorii commentario*); indeed we have a manuscript containing a pictorial description of a counting table named for him (the so-called *Abbonis abacus* – see Oxford, St. John's College, MS 17, fol. 35r, available at WWW: http://digital.library.mcgill.ca/ms-17/folio.php?p=35r). All of these texts were probably created during a quarter of a century, i.e., in a relatively short period between the years 980–1005, and they form the basis (together with other, mostly anonymous tracts, treatises, and commentaries) of a corpus of abacist collections in preserved manuscripts.

72 Cf., for instance, Brown, *The Abacus and the Cross…*, p. 81; or Burnett, "The Abacus at Echternach…," p. 92.

73 See Evans, "Schools and scholars…," p. 71.

Before we move on to the rules of multiplication, it is necessary to clarify some basic information about abacist numeration and explain the essential terms necessary for abacist practice; likewise, Gerbert, in his letters to Constantine in which he discusses an abacus, wrote that it is imperative for people to understand how it is possible for a number to be simple (*simplex*) in one case and composite (*compositus*) in other cases, or to know what *digitus* and *articulus* mean.[74]

However, Gerbert does not explain complex and simple numbers, since, apparently, Constantine already had a sufficient understanding of their meaning. Fortunately, other abacist texts from the 11th century do provide definitions of these terms. An anonymous *Commentary to Gerbert's Regulae* presents them while discussing Gerbert's first rule of multiplication. A simple number (*simplex*) is any number smaller than 10 (*intra denarium*), i.e., numbers 1–9, since, in order to express them, we need only one numerical character (*singularis character*). Given the fact that one numeral is necessary for expressing simple numbers on an abacus, an abacist may consider as simple numbers not only values in order of units but also the numbers 10, 20, 100, 2000, etc. since they require only one symbol located in the appropriate column on the abacus to be expressed.[75]

On the other hand, composite numbers are those which cannot be expressed by one numeral. In this case, two possibilities are available: either both numerals appear simultaneously (*simul*) next to each other, for example, with the value "23" where numeral 2 is placed in the tens column and 3 in the units column, or they are intermitted (*intermissus*), i.e., there are empty columns between the numerals, as in the example of the number "2021," where 2 is placed in the thousands column, the next column to the right is empty, the second 2 is located in the tens column, followed by 1 in the units column.[76]

From these definitions of simple and composite numbers, it seems clear why Gerbert, in his letter to Constantine, insisted that the counting person must understand why a number is sometimes composite and sometimes not. If values 23 and 20 are expressed on an abacus, there is always a numeral 2 in the tens column. However, in the first case, the same numeral is part of a complex number and we must also take into account the numeral in the units column in further counting operations to utilise the value correctly. In contrast, in the second case, the same symbol (i.e., numeral 2) is used as a simple number and, thus, no units

---

74 Gerbert, *Regulae*, praef., p. 7; idem, *Fr. de abaci*, p. 24.
75 *In Gerb.* II, 1, p. 251.
76 Ibid. II, 1, pp. 251–252.

need be considered, and we operate with the 2 in the tens column independently, i.e., differently than with the 2 in the composite number 23.

The differentiation of *digitus* and *articulus* is related to this. *Digiti* correspond to the nine symbols for numbers, i.e., they express the values 1–9 and can be expressed by a single numeral.[77] *Articulus* is necessary when the numerical value is 10 or higher, i.e., the *articulus* itself can be divided into ten identical parts.[78] Thus, in the case of the composite number 23, an abacist considers the numeral 3 to be *digitus* and numeral 2 to be *articulus* – the numeral 2 represents a value of 20, which we can divide into 10 identical parts.

This counting terminology has its origin in the so-called finger-counting that Martianus Cappella mentions concerning arithmetic.[79] *Digitus* (Latin for a finger) is used to show the order of units, since in the process of finger-counting the raised or bent fingers of the left hand were used to represent the values 1–9: i.e., '1' was represented by the little (pinky) finger bent towards the centre of the palm and all other fingers remained raised; '2' was represented by the little finger and ring finger bent in the same manner, while the other fingers were raised; for '3', the middle finger was also bent; '4' was represented by the bent middle finger and ring finger while the remaining three fingers of the left-hand were raised; to represent '5', only the middle finger was bent; '6' was indicated by bending only the ring finger; '7', '8', and '9' were represented similarly to values 1–3, i.e., the bent little finger ('7'), the bent little finger and ring finger ('8'), and the bent little finger, ring finger, and middle finger ('9'), with the difference being that all bent fingers were bent across the whole palm, not towards its centre.[80]

Articulus (Latin for a joint, i.e., the knuckles) gained its name from the fact that the values from 10 to 90 were, via the same method of expressing numerical values, shown by the fingers of the left hand touching the phalanges of specific fingers of the same hand; for example, to express '10', the index finger touched the middle of the thumb; for '30', the thumb and index finger touched via the distal phalanges (or via the nails), while all other fingers were raised; to express '90', the index finger touched the proximal phalanx of the thumb, etc.[81] (see Fig. 36 for a more comprehensive overview[82]).

---

77 Cf. *In Gerb.* II, 2, p. 252.
78 Cf., e.g., *Odon. super abacum*, c. 807A; or Byrhtferth, *De loquela*, c. 688C–D.
79 Martianus, *De nupt.* VII, 729, p. 261.
80 Cf., for example, Bede, *De temp. rat.* 1, pp. 269–270.
81 For more details see, for instance, ibid., pp. 270.
82 See http://www.bl.uk/manuscripts/Viewer.aspx?ref=royal_ms_13_a_xi_f033v [2019-07-17].

Fig. 36 – Expressing numerical values by fingers
(redrawn according to London, British Library, Royal MS 13 A XI, fol. 33v)

Abbo of Fleury, in his commentary to Victorius's *Calculus*, subscribes to this counting tradition when he writes that units are expressed via the left hand in the exact same manner as listed above, whereas in the case of tens, it is necessary to add finger phalanges. Abbo proceeds to the abacist usage of these terms, whereby *digitus* and *articulus* mean parts of a composite number – *digitus* expressing units and *articulus* representing tens. Using the example of the multiplication of tens by tens, he states that in such a case, it is necessary to place the *digitus* of a product into the order of hundreds and the *articulus* into the order of thousands.[83] He accompanies this by an example in which he multiplies 60 by 60, i.e., tens are multiplied by tens while it is enough to multiply 6 · 6 (= 36) and the

---

83   Abbo, *In Calc.* III, 67, p. 115.

digitus of the result is always the lower order of the gained value (i.e., 6) and the *articulus* is the higher order of the given number (i.e., 3). Furthermore, following the introduced rule, it is enough to place the *digitus* of the result (i.e., 6) into the order of hundreds, and the *articulus* (i.e., 3)[84] into the order of thousands. Abbo clearly demonstrates how entrenched the terms *digitus* and *articulus* were – first, they had represented the values 1–9 in the former case, and the values 10, 20, 30,…, 90 in the latter as part of finger-counting; while consequently, they were used for recording values in a decimal system in which *digitus* represented simple (one-digit) numbers, whilst *articulus* expressed tens, although it could be placed in any decimal order.

Bernelin is comparatively brief regarding this issue, only mentioning that if we count using an abacus, it is imperative to know that, using the decimal system, numerical values are sometimes recorded as *digiti* (smaller than ten) and sometimes as *articuli*.[85] Among Gerbert's contemporaries, it was Heriger who commented on this distinction by noting that *digitus* (following what has been said so far) is a number smaller than ten or a simple (*simpliciter*) number, i.e., a number that is added to a higher one, to form a composite number, while *articulus* is the said higher number (i.e., the order of tens), which is followed by a lesser number, i.e., a *digitus* (units).[86]

To sum up, we can say that *digitus* represents numerical values smaller than ten and *articulus* represents numbers that can be divided into ten identical parts, since they are constituted by ten units. By employing the distinction between simple and composite numbers, we may now more easily understand all the following rules for individual counting operations.

The initial challenge for a medieval arithmetician who opted for the decimal positional system was to grasp the new way of recording quantitative values properly. An abacist table made it partly easier for its user by its columnal structure, representing decimal orders, as mentioned above. Although a person of the 21st century will find this system completely natural, it was an absolute novelty a thousand years ago, and it was not easy to understand that values must be recorded from right to left, that it was necessary to take empty columns into account, etc. This stemmed from the fact that the previously common methods of expressing numerical values in words, using fingers or writing them down using

---

84 Ibid. III, 68, p. 115.
85 Bernelin, *Liber abaci* I, p. 26.
86 Heriger, *Regulae*, p. 209; cf. also *In Gerb.* II, 2, p. 252.

Roman (and less commonly Greek) numerals or words, changed entirely when an abacus was used.

The principle of multiplication using an abacus is also readily grasped by us in the 21st century, since it is *de facto* identical with the computing method we use today. However, Gerbert's rules for multiplication might cause some difficulties at first glance. They are introduced as a list of brief regulations which state that when we multiply a specific number in a particular position (column) on the abacus by a number in another (or the same) position on the abacus, then the result should be recorded in a specifically defined column. Once more, it predominantly concerns the recording of numbers using the decimal system. The regulations themselves only show in which order the result of the given multiplication will be placed. All texts originating from the end of the 10th century (including the aforementioned Vigila codex), or soon after the turn of the first millennium contain identical rules, with individual texts differing only in the details of their enumeration.

In the treatise *Regulae de numerorum abaci rationibus,* Gerbert included into his *regulae multiplicationis* 20 (or 21 in the later modified version) rules describing all occurrences in which the numbers located in the first six columns from the right on an abacus are multiplied (i.e., from the order of units to the order of hundreds of thousands – 10^0 to 10^5). In the original version (the *Textus Gerberti genuinus*) we can organise the individual rules into three sequences:

1. nine rules for multiplying factors in the first four columns from the right (i.e., 10^0 to 10^3);
2. six rules for multiplying a factor in the sixth column (10^5);
3. five rules for multiplying a factor in the fifth column (10^4).[87]

By means of this list, Gerbert covered all cases in which numbers ranging from units to hundreds of thousands are multiplied. The only missing rule is for when units are multiplied by units, which is added in the *Textus interpolatus*.[88] When included, the number of regulations in the first sequence thus reaches ten.

A much more detailed list of rules can be found in Heriger's *Regulae numerorum super abacum,* which focuses solely on multiplication. Here, we may find a complete introduction to 181 rules that cover all multiplications that could be computed on an early medieval abacus.[89] The only limitation is presented by

---

87   Gerbert, *Regulae* I, pp. 8–11.
88   Ibid., pp. 8–9.
89   Heriger, *Regula*, pp. 311–324.

the size of the abacus (i.e., 27 columns) which allowed for arithmetical operations with integers up to 10^{26}. This means that the highest multiplied numbers could have been those whose product would not have exceeded hundreds of septillions. Obviously, the infinite numerical series can proceed to higher orders, as Heriger knew very well;[90] however, a twenty-seven-column abacus cannot be used for these multiplications.

The infinite numerical series is mentioned in the introduction to the so-called *Abbonis abacus*, which can serve as evidence for the claim that prescholastic or early scholastic abacists differentiated between the mathematical series themselves and the possibilities offered by their computing tool.[91] In his commentary to Victorius, Abbo mentions only selected cases of multiplying individual orders (14 rules in total),[92] which resembles Gerbert's list. Bernelin is comparatively brief in his *Liber abaci*, offering eight rules in total (which are also found in Gerbert's texts).[93] To this brief exposition on placing the results of multiplied factors in individual columns, which has a primarily illustrative character, Bernelin adds that the correct placement of *digitus* and *articulus* of products is the same, provided that the factors do not change by one order – therefore, both the *digitus* and *articulus* of the result are placed identically for multiplying units by thousands, for multiplying tens by hundreds, for multiplying units by millions, and for multiplying tens by hundreds of thousands.[94]

The relatively brief and monotonous enumeration of rules provided by Gerbert (both in the commentaries as part of *Textus interpolatus* and in other contemporary texts) is accompanied by another simple aid for eager arithmeticians. Apparently, not every abacus user was able to memorise all the rules about placing *digiti* and *articuli* for individual orders of factors; therefore, a general notice about the placement of a product for a given order was added. In the case of multiplying units, the *digitus* is always written in the same column as the multiplier, the *articulus* then shifts one column to the left; if we multiply by tens, then the *digitus* is placed in the second column to the left of the multiplier and the *articulus* moves to the third column (i.e., a shift by one and two orders respectively); for a multiplicand in the order of hundreds, the *digitus* is placed in the third column to the left of the multiplier and the *articulus* moves into the fourth column (i.e., a shift by two and three orders respectively); when thousands are

---

90   Heriger, *Regulae*, p. 209.
91   Abbo, *Abacus*, p. 203.
92   Abbo, *In Calc.* III, 64, pp. 113–114.
93   Bernelin, *Liber abaci* I, p. 27.
94   Ibid.

multiplied, the *digitus* and *articiulus* are always placed in the fourth and fifth column from a multiplicand (i.e., a shift by three and four orders respectively); for further orders, the *digitus* of a result with a multiplicand in a higher order occupies the place of the result of the *articulus* from the preceding order.[95]

Thus, essentially all the requisites of specific or general rules for multiplication using an abacus are covered. A would-be arithmetician learned a single piece of information – if he multiplied by one particular numerical order, the *digitus* of the result should occupy a specific order and the *articulus* should be placed in the closest higher order.

It is evident that if anybody wanted to use an abacus, it was necessary to possess other knowledge, especially an understanding of the so-called multiplication table. Gerbert (as well as Heriger) does not cover this issue, since he knew that the addressee of his treatise (Constantine of Fleury) was well versed in this art. By the end of the 10th century, arithmeticians were able to use the aforementioned *Calculus* by Victorius of Aquitaine[96] or Boethius's *Introduction to arithmetic*.[97]

However, we can also find a multiplication table overview in an abacist text originating around the year 1000 – the table, with 36 cases of multiplying, is included in Bernelin's *Liber abaci*. Knowledge of this table would have facilitated computing on an abacus and highlighted the usefulness of this computing aid. It is a multiplication of *digiti* (i.e., a multiplication table) and, according to preserved manuscripts, Bernelin either omits cases in which *digiti* multiply themselves (i.e., square roots)[98] or he omits multiplicands of one (except for its square root),[99] while he takes into account the commutativity of factors (i.e., the identical results regardless of whether we multiply $a \cdot b$ or $b \cdot a$; thus he does not repeat cases with switched order of factors). For a comprehensive overview, see Table 15.[100]

---

95   Gerbert, *Regulae* I, pp. 10–11.
96   Victorius created a collection of forty-nine tables that provide results of multiplying by two to fifty for values from 1000 to 1/144 – see Victorius, *Calc.*, pp. 4–37.
97   Cf., especially, Boethius, *Arith.* I, 26, p. 64.
98   See London, British Library, Add MS 17808, fols. 57v–58r (http://www.bl.uk/manuscripts/Viewer.aspx?ref=add_ms_17808 [2020-03-18]).
99   Cf. München, Bayerische Staatsbibliothek, Clm 14689, fol. 50r (https://daten.digitale-sammlungen.de/~db/0004/bsb00041143/images/index.html?id=00041143&nativeno=50 [2020-03-18]).
100 Cf. Bernelin, *Liber abaci* I, pp. 25–26. Values in square brackets represent the cases missing in some manuscripts.

Table 15 – Multiplication table according to Bernelin's *Liber abaci*

								[9·9=81]
							[8·8=64]	8·9=72
						[7·7=49]	7·8=56	7·9=63
					[6·6=36]	6·7=42	6·8=48	6·6=54
				[5·5=25]	5·6=30	5·7=35	5·8=40	5·9=45
			[4·4=16]	4·5=20	4·6=24	4·7=28	4·8=32	4·9=36
		[3·3=9]	3·4=12	3·5=15	3·6=18	3·7=21	3·8=24	3·9=27
	[2·2=4]	2·3=6	2·4=8	2·5=10	2·6=12	2·7=14	2·8=16	2·9=18
[1·1=1]	[1·2=2]	[1·3=3]	[1·4=4]	[1·5=5]	[1·6=6]	[1·7=7]	[1·8=8]	[1·9=9]

The reason for omitting cases of a higher number multiplied by a smaller number is easily explained – according to abacists, a lower number was always to be multiplied by a higher one and not *vice versa*. Concurrently, as stated by Bernelin, it is possible to combine factors mutually. Thus, it is not a problem to order numbers on an abacus so that the multiplicand is lower than the multiplier.[101] The absence of square roots is more surprising. Perhaps Bernelin strictly adhered to the rule that the multiplicand is always smaller; therefore, multiplication cases such as 2 · 2, 5 · 5, etc. are not included. Alternatively, he could have considered these cases so obvious that it was unnecessary to list them, or the addressee of the treatise, the otherwise unknown Amelius,[102] was aware of the table for multiplying a number with an identical one, presented by Victorius of Aquitaine.[103]

Nevertheless, with knowledge of the multiplication table, the multiplication process on an abacus is quite straightforward. Although specific examples are missing from Gerbert's (and Heriger's) account, Bernelin (who otherwise does not focus on rules for placing *digiti* and *articuli* of multiplication results) presents a comparatively complex example by which he demonstrates that multiplication using an early medieval computing table is very simple (*facilius*).[104] On completion of the task, we should learn what totals we will get if we imagine a tower (*turris*) that has 12 windows (*fenestrae*); in each of them, there are 12 rugs (*stratus*); on each of which stand 12 men (*viri*); each man has 12 wives (*mulieres*); and each wife nurses 12 children (*infantes*). It is a sum of square roots, i.e., multiplications. Thus, it is necessary to compute the values 12^2 (i.e., 12 · 12), 12^3 (i.e., 12 · 12 · 12), 12^4 (i.e., 12 · 12 · 12 · 12) and 12^5 (i.e., 12 · 12 · 12 · 12 · 12).

In order to produce a result, it is necessary to place tokens with numerical symbols accordingly, to know the multiplication table, and to move the products of individual multiplications to the corresponding orders of factors. A necessary prerequisite is also a knowledge of summation. Bernelin describes the procedure as follows (cf. Fig. 37):

I. placement 1 (= tower);
II. placement 12 (= number of windows)

---

101 Bernelin, *Liber abaci* I, p. 27.
102 Ibid., p. 16.
103 See Victorius, *Calc.*, p. 48.
104 Bernelin, *Liber abaci* I, p. 27.

III. multiplication 12 · 12, meaning:
   a) $2 \cdot 2 = 4$; 4 in the order of units;
   b) $2 \cdot 1[0] = 2[0]$; 2 in the order of tens;
   c) $1[0] \cdot 2 = 2[0]$; 2 in the order of tens;
   d) $1[0] \cdot 1[0] = 1[00]$; 1 in the order of hundreds;
   e) sum of gained products: $4 + 2[0] + 2[0] + 1[00] = 144$ (= number of rugs);
IV. multiplication of gained product by twelve, i.e., $144 \cdot 12$, meaning:
   a) $2 \cdot 4 = 8$; 8 in the order of units;
   b) $2 \cdot 4[0] = 8[0]$; 8 in the order of tens;
   c) $2 \cdot 1[00] = 2[00]$; 2 in the order of hundreds;
   d) $1[0] \cdot 4 = 4[0]$; 4 in the order of tens;
   e) $1[0] \cdot 4[0] = 4[00]$; 4 in the order of hundreds;
   f) $1[0] \cdot 1[00] = 1[000]$; 1 in the order of thousands;
   g) sum of results: $8 + 8[0] + 4[0] + 2[00] + 4[00] + 1[000] = 1728$ (= number of men);
V. multiplying second product by twelve, i.e., $1728 \cdot 12$, meaning:
   a) $2 \cdot 8 = 16$;
   b) $2 \cdot 2[0] = 4[0]$;
   c) $2 \cdot 7[00] = 14[00]$;
   d) $2 \cdot 1[000] = 2[000]$;
   e) $1[0] \cdot 8 = 8[0]$;
   f) $1[0] \cdot 2[0] = 2[00]$;
   g) $1[0] \cdot 7[00] = 7[000]$;
   h) $1[0] \cdot 1[000] = 1[0000]$;
   i) sum of results: $16 + 4[0] + 8[0] + 2[00] + 14[00] + 2[000] + 7[000] + 1[0000] = 2[0]736$ (= number of wives);
VI. multiplying third product by twelve, i.e., $2[0]736 \cdot 12$, meaning:
   a) $2 \cdot 6 = 12$;
   b) $2 \cdot 3[0] = 6[0]$;
   c) $2 \cdot 7[00] = 14[00]$;
   d) $2 \cdot 2[0000] = 4[0000]$;
   e) $1[0] \cdot 6 = 6[0]$;
   f) $1[0] \cdot 3[0] = 3[00]$;
   g) $1[0] \cdot 7[00] = 7[000]$;
   h) $1[0] \cdot 2[0000] = 2[00000]$;
   i) Sum of results: $12 + 6[0] + 6[0] + 3[00] + 14[00] + 7[000] + 4[0000] + 2[00000] = 248\,832$ (= number of children);
VII. final sum: $1 + 12 + 144 + 1728 + 2[0]736 + 248\,832 = 271\,453$ (= final result).

$1 + 12 + 12^2 + 12^3 + 12^4 + 12^5 = 271456$

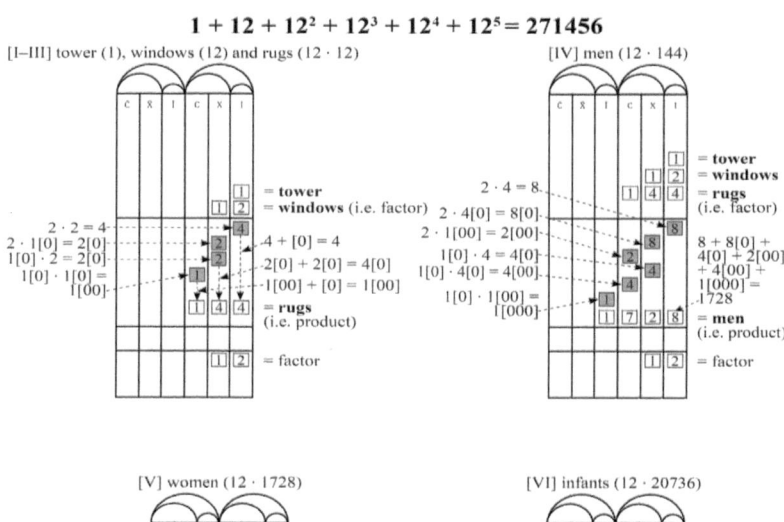

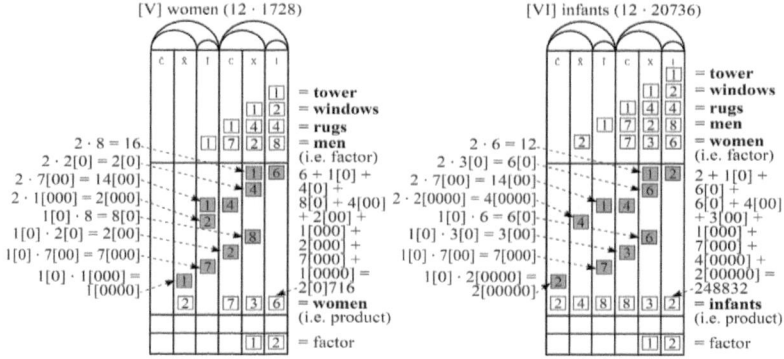

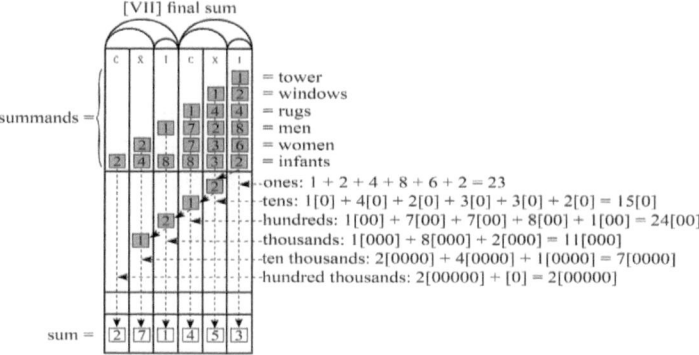

Fig. 37 – Multiplication according to Bernelin's *Liber abaci* I

With a knowledge of the table of multiplication and the rules for placing the *digiti* and *articuli* of a product, the process of multiplication on an abacus becomes a rather elementary mathematical operation that is entirely intuitive for a person of the 21st century.

3. *Regulae divisionis*

The highest art of every arithmetician is to master division; therefore, many medieval abacist handbooks, including those by Gerbert himself, devoted the majority of space to this arithmetical operation. Regarding division using an abacus, a maximal effort was made to reduce the division itself by converting individual operations to multiplying or subtracting and adding. Additionally, both divisor and dividend are very often modified so that the values are as easy to compute as possible. Gerbert's *Regulae* are relatively brief, even on this point. Only a few steps are fully explained and justified; thus, it would be beneficial to focus first on abacist division in a broader context, although Gerbert's perspective remains our primary concern.

If appropriate, an abacist may decide to adjust a divisor by using a so-called difference (*differentia*), which is known simply as 'division with a difference' (*divisio cum differentia*).[105] In the case of division with a difference, a value is added to a divisor, allowing a more straightforward calculation of the quotient. For example, dividing by eight is undoubtedly more complicated than dividing by ten; therefore, a value of two is added to the divisor, and the divisor of ten is used for the computation (to determine the remainder of the division, the multiples of this difference by the individual quotients are added to the remainder of the dividend).[106]

Cases in which the divisor is a composite number are processed similarly. Two options are available – either the difference is adjusted so that the sum of the difference and the divisor produces a number whose highest numeral moves to the closest higher order and is a one (e.g., to the divisor 56, we add the difference 44, and the divisor with a difference becomes 100); or a difference is chosen so that it gives a number with zeroes (empty spaces on an abacus) in lower orders (e.g., using the same divisor 56, we may add the difference 4, and the divisor becomes 60). In this latter case, abacists say that one increases the highest numeral of the divisor, determining which part of the dividend will be detached, i.e., for the given example (56 + 4), it will be a sixth of the dividend.[107]

---

105 Cf., for example, Bernelin, *Liber abaci* II, pp. 40–41.
106 Cf., e.g., Gerbert, *Regulae* II, 1 [2], p. 13.
107 Cf. ibid. II, 2 [4], pp. 14–15; or Bernelin, *Liber abaci* III, pp. 50–51.

If we add a difference to the divisor so that the result is a one in the highest order of the divisor (i.e., the divisor will be 10, 100, 1000, etc.), it is known as 'simple division' (*diviso simplex*),[108] since the division is made by a simple divisor (*divisor simplex*).[109] In the case of a composite divisor (*divisor compositus*)[110] when there is a number other than one in the highest order (i.e., for example, 60, 500, 4000, etc.), it is called a 'composite division' (*divisio composita*).[111] For simple division, it is sufficient to move the highest numeral of the dividend on an abacus into the results according to the placement of the divisor's one by a corresponding number of columns: i.e., a divisor of 10 will reduce the order of the dividend by one, a divisor of 100 will reduce the dividend by two orders, etc.[112] For a composite division, the appropriate part of the highest number of the dividend is determined by the value of the highest number of the divisor with a difference, or, in other words, according to the value by which the difference increased the divisor, i.e., with a divisor of 17, the divisor with a difference is 20; thus a half of the dividend will be determined, whereas a divisor of 38 means it will be a quarter, since the divisor with a difference is 40, etc.[113]

During the process of composite division, the abacist also tries to adjust the dividend, i.e., to modify it in such a way that the highest number of the modified dividend is one, e.g., if the dividend is the number 8456, then the abacist excludes one thousand (1[000]) from its numeral in the highest order (i.e., 8[000]), thus creating a so-called reduced dividend. According to the value of the divisor, this reduced dividend might be further reduced so that the computations can be made in the lowest orders possible, e.g., if the number 38 is the divisor, determining that a quarter of the dividend is sought, then the highest number of the divisor in the order of tens allows for a reduction in the value of the mentioned reduced dividend by one order, i.e., the current dividend in the process of a so-called reduced division will be a hundred.[114] Quotients gained through the division of a divisor reduced in this way are understood as reduced results. Their value is used to multiply the highest number of the original (or the current for further steps) divisor; these products then provide partial results.[115]

---

108 Bernelin, *Liber abaci* II, pp. 31–48.
109 Gerbert, *Regulae* II, 4 [7], p. 17.
110 Ibid.
111 Bernelin, *Liber abaci* III, pp. 50–73.
112 Cf. Gerbert, *Regulae* II, 1 [2–3], pp. 12–14; or Bernelin, *Liber abaci* II, pp. 33–35.
113 Cf. Gerbert, *Regulae* II, 2 [4–6], pp. 14–17; or Bernelin, *Liber abaci* III, pp. 53–55.
114 Cf. Gerbert, *Regulae* II, 2 [5], pp. 15–16.
115 Cf. ibid., p. 16.

Presenting model examples that follow Gerbert's individual rules for division will perhaps make this more comprehensible. *Textus genuinus* presents six rules for division, while *Textus interpolatus* increases their number to ten. Most editors and researchers who have examined Gerbert's division according to his *Regulae* during the last centuries (see A. Olleris,[116] M. Chasles,[117] G. Friedlein,[118] similarly N. Bubnov) have followed the ordering and wording of rules from the expanded version, and we will respect this approach here. However, for comparison, it might be helpful to present a table comparing the ordering of rules for division in editions of Gerbert's *Regulae* – see Table 16.

Table 16 – Different orderings of Gerbert's rules of division

Bubnov – textus genuinus	Bubnov – textus interpolatus	Olleris	Chasles	Friedlein	examples
3	1	Cap. VI.	I.	I.	7 : 3, 9[0] : 2[0]
1	2	Cap. VII.	II.	II.	896 : 7, 61 : 7
---	3	Cap. VIII.	III.	III.	700 : 80
2	4	Cap. IX.	IV.	IV.	97 : 22, 124 : 17
---	5	Cap. X.	V.	V.	724 : 17
---	6	Cap. XI.	VI.	VI.	700 : 280
4	7	Cap. XII.	VII.	VII.	8000 : 600, 8000 : 375
5	8	Cap. XIII.	VIII.	VIII.	900 : 204
---	9 (+ 9 bis)	Cap. XIV.	VIIII.	IX.	8000 : 207, 40000 : 3009
6	10	Cap. XV.	X.	X.	all rules

The first rule of *Textus interpolatus* explains the process of division in cases when both dividend and divisor are in the same order (i.e., 9 : 2 or 70 : 30, etc.). The text is very brief, stating that units are always subtracted (*subtrahere*) from units, while these units correspond to *digiti* in any order, since, for example, 90 : 20 is essentially 9 : 2, even though the symbols of numerals are placed in the order of tens on an abacus. Effectively, we subtract the divisor from the dividend as many times as possible. The number of subtractions corresponds to the

---

116 Idem, *De num. div.*, pp. 351–355.
117 Idem, *Traité*, pp. 297–299.
118 See Friedlein, G. "Gerbert's Regeln der Division." *Zeitschrift für Mathematik und Physik* 9 (1864), pp. 145–163.

result of the division. In other words, we find out how many times the divisor is included in the dividend.[119] Therefore, we can say that the computation of the example 90 : 20 might have looked like this (see Fig. 38-I):

a) first subtraction of the divisor from the dividend, i.e., 9[0] − 2[0] = 7[0];
b) second subtraction of the divisor from the new dividend, i.e., 7[0] − 2[0] = 5[0];
c) third subtraction of the divisor from the new dividend, i.e., 5[0] − 2[0] = 3[0];
d) fourth subtraction of the divisor from the new dividend, i.e., 3[0] − 2[0] = 1[0];
e) the divisor was subtracted four times; thus, the result of division 9[0]: 2[0] = 4; the remaining part of the divisor is 1[0], which is the remainder of the division.

Alternatively, Gerbert might have referenced this procedure (see Fig. 38-II):

a) divide in the order of tens (how many times the divisor is contained in the dividend), i.e., 9[0]: 2[0] = 4, which is the result;
b) multiply the gained result (4) by the divisor (2[0]), i.e., 4 · 2[0] = 8[0];
c) subtract the gained product from the dividend, i.e., 9[0] − 8[0] = 1[0], which is the remainder from the division;
d) thus 9[0]: 2[0] = 4, and the remainder is 1[0].

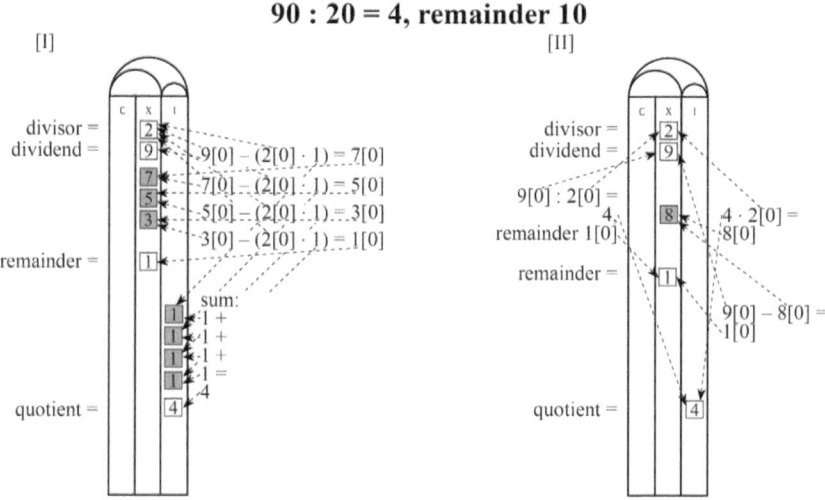

Fig. 38 – Division according to Gerbert's 1st rule

---

119 Gerbert, *Regulae* II, 3 [1], pp. 12, 15.

The second rule of division explains the procedure of simple division with a difference, where the divisor is primarily in the order of units (similarly, in the order of tens and higher orders) and the dividend is in a higher order (tens, hundreds, thousands, etc.), e.g., 896 : 7. To the divisor (*divisor*) in the order of units (*singularis*), a difference (*differentia*) up to ten is added, e.g., in the case of the divisor 7, the divisor with a difference is 10, so the division is done by *articulus*. If the divisor with a difference has 1 in the order of tens, it is clear that the highest numeral of the dividend is moved one order lower in the partial results (tens are reduced to units, hundreds to tens, etc.). Then it is necessary to multiply (*multiplicare*) the partial result gained this way by the difference and to add it to the remainder from the divisor. The same procedure is followed until the remainder of the dividend becomes smaller than the divisor.[120] An example of this could be the division 61 : 7, which, following Gebert's rule, might have looked like this (see Fig. 39):

a) determine the difference, i.e., 7 + 3 = 1[0];
b) first division, i.e., 6[0]: 1[0] = 6, the 6 from the dividend is moved to the order of units as the first partial result;
c) multiply the first partial result (6) by the difference (3), i.e., 6 · 3 = 18, meaning that the *articulus* is placed in the order of tens and the *digitus* in the order of units as the new dividend;
d) second division, i.e., 1[0]: 1[0] = 1, the 1 from the dividend is moved to the order of units as the second partial result;
e) multiply the second partial result (1) by the difference (3), i.e., 1 · 3 = 3, the 3 is placed in the order of units for the divisor;
f) the sum of the units of the dividend, i.e., 1 + 8 + 3 = 12; thus, the new dividend has an *articulus* in the order of tens and a *digitus* in the order of units;
g) third division, i.e., 1[0]: 1[0] = 1, the 1 from the dividend is moved to the order of units as the third partial result;
h) multiply the third partial result (1) by the difference (3), i.e., 1 · 3 = 3, the 3 is added to the dividend in the order of units;
i) the sum of the units of the dividend, i.e., 2 + 3 = 5, the sum is smaller than the divisor; thus, this is the remainder of the entire division;
j) the sum of the partial results, i.e., 6 + 1 + 1 = 8, the result of the division 61 : 7 is thus 8, remainder 5.

---

120 Ibid. II, 1 [2], pp. 12–13.

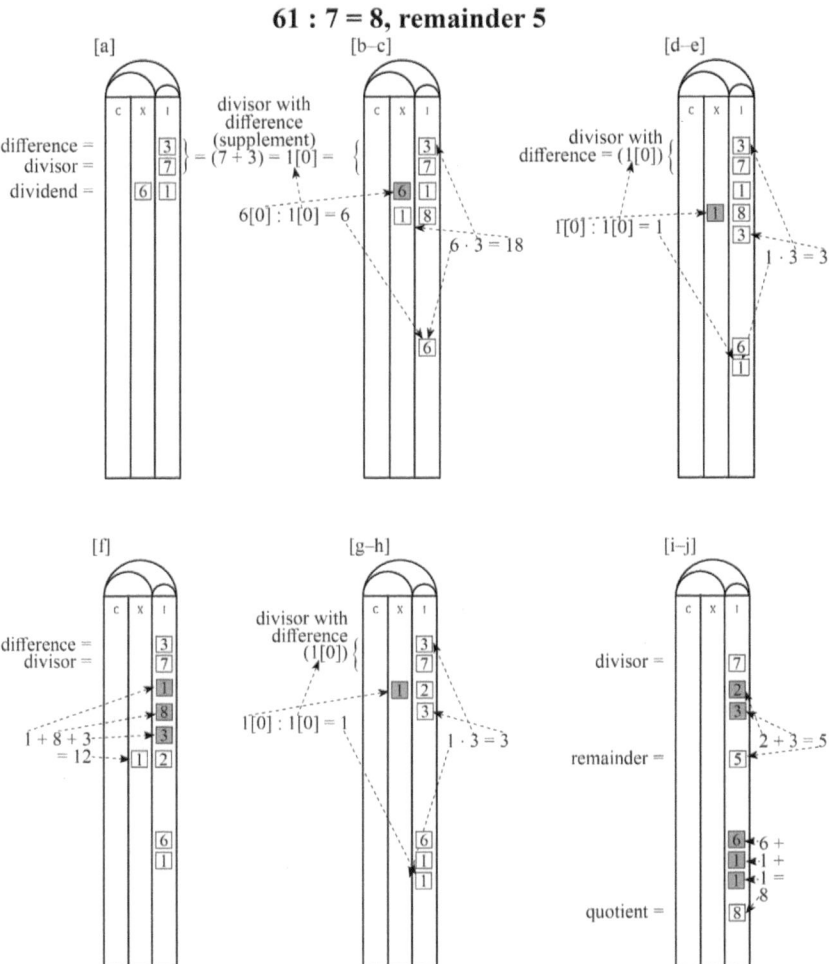

Fig. 39 – Division according to Gerbert's 2nd rule

The following rule is likewise an example of simple division with a difference; however, the divisor is placed in the orders of tens, hundreds, etc. (meaning that there is nothing in the order of units, i.e., there is a zero) and the dividend is a number at least one order higher. Here, the procedure is the same as in the case of the second rule. Only the *digitus* of the divisor in the order of tens and higher (i.e., the divisor, e.g., 5[0], 7[00], etc.) is thought of as a unit (*quasi singularis*) – the difference is added to it and different columns of the abacus are employed in the same manner as in the second rule.[121]

The fourth and the fifth rule represent procedures for composite division with a difference in which the divisor is a number composed of a *digitus* and an *articulus* in the orders of units and tens, while the dividend is a simple number (e.g., 1[00]) or a composite number in the orders of tens, hundreds, thousands, etc., regardless of whether there are numerals in every order (e.g., number 12) or an order is empty (e.g., 1[0]5). The difference for the divisor composed of units and tens (e.g., 26) is determined in such a way that the column of units remains empty (i.e., 26 + 4 = 3[0]). Gerbert writes that an arithmetician must realise in which tens we may find the unit of such a divisor.[122] In the listed example, it is clear that the unit is in the third tens (*tertius singularis*), which means that the divisor is composed of the *articulus* 2[0] and the *digitus* 6; therefore, all numbers of the third ten are composed of twenty (*viginti*) and a unit.[123] In other words, the second ten includes numbers 11–19, the third ten 21–29, etc.

The awareness of the ten in which a unit of a composed divisor is located is crucial for the delineated divisors with which the abacist will operate. If a unit of the divisor is in the second ten, then this two determines that half of the dividend will be detached. If a unit is in the third ten, one-third of the dividend will be detached, etc. Contrary to simple division, in which the highest numeral of the dividend is moved to partial results in a lower order according to the value of the divisor with a difference (i.e., ten reduces a result by one order, a hundred by two), in the case of a composite divisor, the highest numeral of the divisor is not moved entirely but only its relative part (one half, one third, one fourth, etc.) in accordance with the ten in which the unit of the divisor composed of units and tens is located. The further procedure is the same as with the previous two rules: a partial result is multiplied by the difference and then added to the remainder from the division of the dividend; the process is repeated until the remaining part of the dividend is smaller than the divisor. The principle of composite division with a difference is more comprehensive in the example 97 : 28 (see Fig. 40):

a) determine the difference, i.e., 28 + 2 = 3[0]; the difference is 2, meaning that the divisor with a difference is 3[0], which means that the relative part of the dividend is one third;

---

121 Ibid. II, 1 [3], pp. 13–14.
122 Ibid. II, 2 [4], p. 14.
123 Friedlein, G., *Die Zahlzeichen und das elementare Rechnen der Griechen un Römer und des christlichen Abendlandes vom 7. bis 13. Jahrhundert*. Erlangen: A. Deichert, 1869, p. 107.

b) from the highest numeral of the dividend (9[0]), detach the relative part, i.e., one third, 9[0]: 3[0] = 3; place the 3 into the order of units as the first partial result;
c) multiply the first partial result by the difference, i.e., 3 · 2 = 6; place the 6 in the order of units for the dividend;
d) find the sum of the *digitus* of the original dividend (7) and the gained product (6), i.e., 7 + 6 = 13; which gives a value smaller than the divisor and constitutes the remainder of the division;
e) thus 97 : 28 = 3, remainder 13.

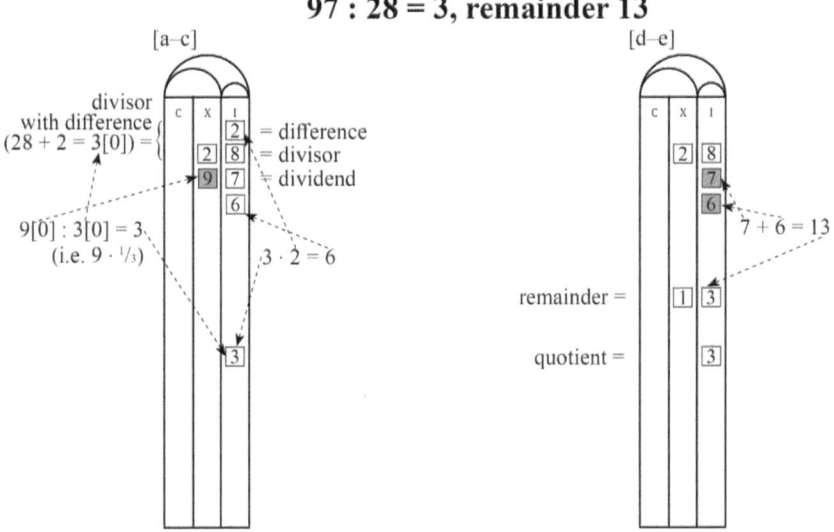

Fig. 40 – Division according to Gerbert's 4th rule

The second part of the fourth rule and the fifth rule further specify the procedure of division in cases where the highest numeral of the dividend is located in a higher order than the highest numeral of the divisor. In such cases, Gerbert recommends decomposition (*dissipare*) to make the process of division easier. First, it is necessary to detach a unit from the highest numeral of the dividend – i.e., to reduce the dividend – e.g., if the number 7568 is the dividend, then from 7[000] we detach 1[000], which we can further reduce by one order to 1[00]. This makes it easier to search for the appropriate part of the dividend according to the unit of the divisor (which is still composed of units and tens, e.g., 28; thus, we know to search for a third of the divisor). In the presented example, one-third of ten is sought, since the detached thousand was reduced to a hundred and the *digitus* of the divisor is in the third ten; thus, the order of units can be omitted and we get 10 · 1/3 or 10 : 3.[124] The further procedure is then similar to that mentioned above. However, it is necessary to consider the fact that the divisor was reduced to a hundred, thousand, etc.; thus, it is imperative to multiply the gained partial results and products of the partial results with a difference, which are added to the dividend by the original proposed (*proposita*) value. A demonstration is provided by the example of 724 : 17 (see Fig. 41):

a) the unit in the second ten determines that the relative part of the dividend will be one half;
b) set the difference of the divisor, i.e., 17 + 3 = 2[0]; the difference is thus 3;
c) from the highest numeral of the dividend (7[00]), detach a hundred (1[00] = reduced dividend);
d) the reduced dividend is broken down into tens, i.e., 1[00] : 1[0] = 1[0];
e) the broken-down reduced dividend (1[0]) is split in half, i.e., 1[0] : 2 = 5; 5 is placed in the order of units as the reduced result;
f) the reduced result (5) is multiplied by the difference (3), i.e., 5 · 3 = 15; 1 is placed in tens and five into units as the reduced remainder;
g) the reduced result (5) is multiplied by the highest number of the original dividend (7), i.e., 5 · 7 = 35; 3 is placed into tens and five into units as the first partial result;
h) the reduced remainder (15) is multiplied by the highest numeral of the dividend (7), i.e., 15 · 7 = 1[0]5; 1 is placed into hundreds, and 5 into units as the new dividend;
i) the reduced result (5) is multiplied by the highest numeral of the new dividend (1), i.e., 5 · 1 = 5; 5 is placed into units as the second partial result;
j) the reduced remainder (15) is multiplied by the highest numeral of the new dividend (1), i.e., 15 · 1 = 15; 1 is placed next to the dividend into tens, and 5 into units;
k) add tens and units of the original dividend (24) to the units of the new dividend (5) and to the value of the last product (15), i.e., 24 + 5 + 15 = 44; thus, a new dividend is gained whose highest numeral is in the order of tens; therefore, it is not broken down further;

---

124 Gerbert, *Regulae* II, 2 [4–5], pp. 15–16. Cf. Chasles, M., "Analyse et explication du traité de Gerbert." *Comptes Rendus des séances de l'Académie des Sciences* 16 (1843), pp. 288–289; or Friedlein, "Gerbert's Regeln…," p. 154.

l) subtract the divisor (without the difference) from the new dividend (44) as many times as possible, i.e., 44 − 17 = 27, 27 − 17 = 10, which means twice; therefore, place 2 in units as a partial result; the remaining 1 in tens constitutes the remainder of the entire division;
m) add together the partial results 35 + 5 + 2 = 42, which is the result; thus, 724 : 17 = 42, remainder 10.

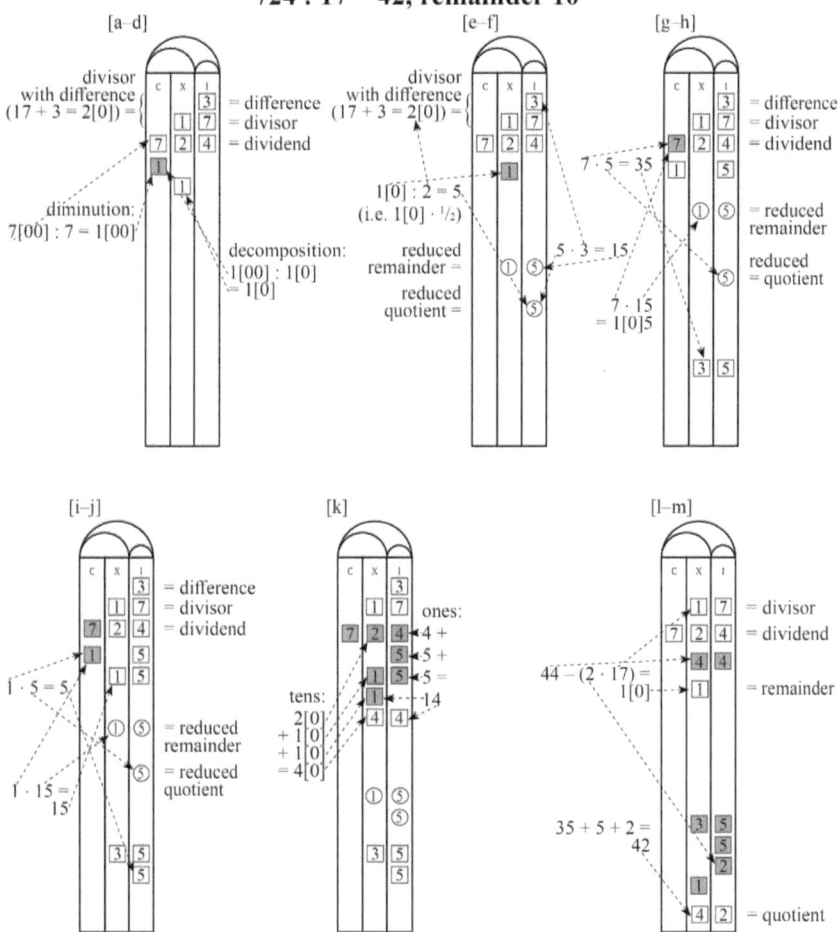

Fig. 41 – Division according to Gerbert's 5th rule

The sixth rule is analogical to the previous two rules, since it increases the composite divisor and dividend by at least one order. Thus, the divisor could be, for example, the number 280, while the dividend could be the numbers 800 or 6456, etc. The process is the same as with the earlier rules for composite division with a difference: it is necessary to find out which part of the composite number of the divisor is constituted by the *digitus* and, consequently, to search for the corresponding part of the dividend, multiply the partial results by the difference, etc.[125] Similarly, the seventh rule sums up all existing division regulations so far and allows for the use of the same procedure for simple and composite division in any orders.[126]

Apart from division with a difference, which was later called 'iron division' (*divisio ferrea*),[127] an abacist can also employ computing without a difference (*sine differentia*).[128] Thus, an arithmetician uses the divisor in its original form and does not add any difference that would facilitate the process of computing. Later texts called this kind of division a 'golden division' (*divisio aurea*)[129] and we can find it in Gerbert's *Regulae* as well. In *Liber abaci*, Bernelin likens division without a difference to a lady or a mistress (*domina*), since it is not necessary to add a difference and its division always takes place, while division with a difference is likened to a maid (*famula*) since it necessitates a servant in the form of the difference.[130]

The method of division without a difference largely resembles composite division with a difference; however (since no difference is added to the divisor) the value of the divisor is not increased by one in the highest order of tens; thus, the highest numeral of the divisor determines the relative part of the dividend that will be detached (e.g., if the number 34 is the divisor, a third of the dividend is sought). In order to account for numerals of the divisor in lower orders (*minuti*), it is necessary to be sure that there is something from which we can subtract products of partial results with values in lower orders of the divisor. This requires certain modifications of the dividend, from which a specific part is detached (e.g., a hundred, thousand, etc., i.e., a unit from the highest numeral of the dividend). The process of division without a difference by a composite divisor bears a resemblance to Gerbert's composite division with a difference; however, he firstly decomposes the dividend. Regarding the eighth rule, Gerbert lists examples of

---

125 Gerbert, *Regulae* II, 2 [6], pp. 16–17.
126 Ibid. II, 4 [7], p. 17.
127 See, for example, Adelard, *Reg. abaci*, p. 91.
128 Bernelin, *Liber abaci* II, p. 41.
129 Cf., e.g.,. např. Adelard, *Reg. abaci*, p. 91.
130 Bernelin, *Liber abaci* II, p. 41.

division in which the highest numeral of both dividend and divisor is of the same order, whilst the divisor has at least one empty column (i.e., there is a zero), e.g., dividing 800 by 305. As for the ninth rule, the highest numeral of the dividend is located in a higher order than the highest numeral of the divisor, e.g., 8000 : 305.[131] Let us look at an illustrative example – to divide 40 000 by 3009, we may follow this procedure (cf. Fig. 42):

a) the dividend is reduced to ten thousand, i.e., 4[0000]: 4 = 1[0000];
b) the reduced dividend is decomposed to thousands, i.e., 1[0000] = 9[000] + 1[000];
c) the higher part of the decomposed reduced divisor (9[000]) is compared to the highest numeral of the divisor (3), meaning that we determine its third, i.e., 9[000]: 3[000] = 3; the 3 in the order of units becomes the reduced result;
d) by the reduced result (3), we multiply a number in a lower order of the divisor (9), i.e., 3 · 9 = 27, which we subtract from the thousand detached in step b), i.e., 1[000] – 27 = 973, which constitutes the reduced remainder;
e) by the reduced result (3), we multiply the highest numeral of the original dividend (4), i.e., 3 · 4 = 12; thus, we gain the first partial result;
f) by the reduced remainder (973), we multiply the highest numeral of the original dividend (4), i.e., 973 · 4 = 3892; thus, we gain a new dividend whose highest numeral is in the same order as the highest numeral of the divisor;
g) we compare the new dividend (3892) and divisor (3[00]9), meaning: 3892 – 3[00]9 = 883; thus, the divisor is contained by the dividend once – an one in the order of units is the second partial result and 883 is the remainder of the entire division;
h) we add together the partial results, i.e., 12 + 1 = 13, which is the result; thus 4[0000]: 3[00]9 = 13, remainder 883.

---

131 Gerbert, *Regulae* II, 5 [8–9], pp. 18–21.

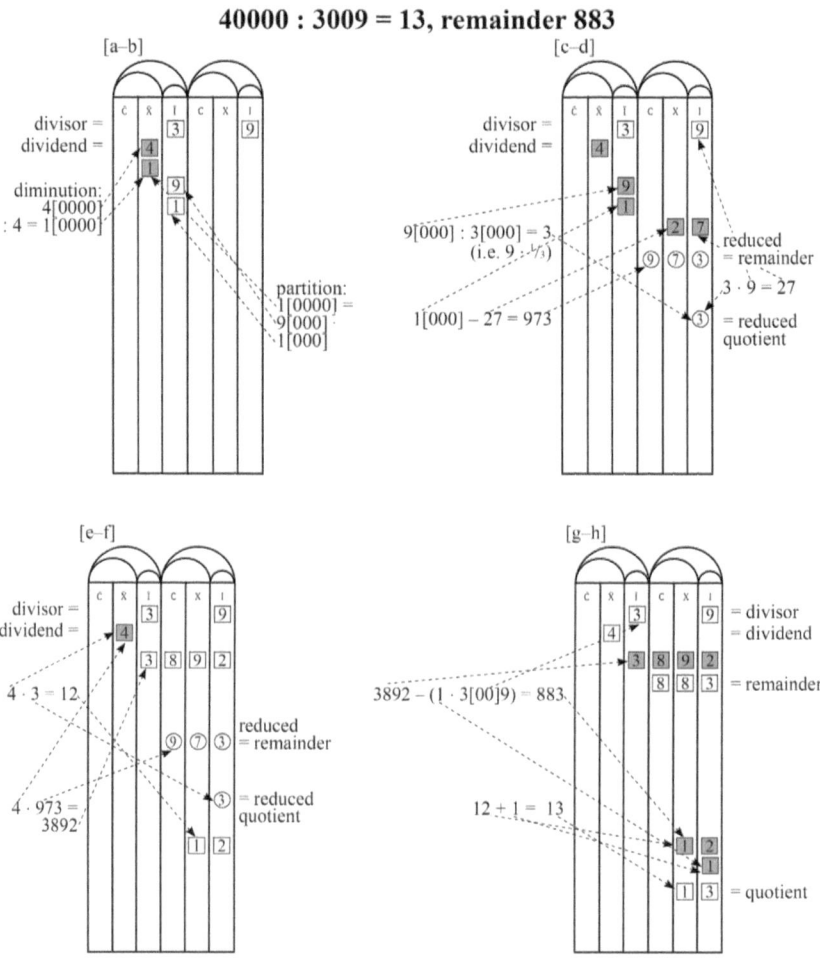

Fig. 42 – Division according to Gerbert's 9th rule

Thus, Gerbert covered all the examples of division on an abacus – both simple and composite, with a difference and without. At the end, he adds one last rule, which summarises and repeats how to establish the number of times the divisor is contained in any dividend. It is reiterated that should any highest numeral both of dividend and divisor be located in the same order (e.g., rule 1), then the divisor is subtracted from the dividend and the number of possible subtractions corresponds to the number of divisors included in the dividend (e.g., 8 : 2 = 4 since two can be subtracted from eight four times). Therefore, we add together (*aggregare*) how many times we can subtract the divisor from the dividend, since dividends constitute collections (*collectiones*) of these divisors.[132]

In the case of simple division with a difference (i.e., the divisor with a difference is a ten, hundred, etc., cf. rules 2 and 3), the number of divisors in the dividend is determined by shifting the highest integer of the dividend by a corresponding number of columns to the partial results (e.g., 2[00]: 1[0] = 2[0], i.e., the 2 from the dividend is shifted by one order lower to pose as the result) since the result is derived from a whole (*a toto*) since the divisor is a 1 in any order.

In the case of composite division with a difference, the highest integer of the dividend is not moved to the result but only its part (see rule 4), which is specified by which ten of the divisor contains the unit (e.g., if the divisor is number 36, the unit of the divisor, i.e., 6, is located in the fourth ten; thus, one fourth of the dividend will be detached). Therefore, the ten that contains the divisor unit determines which part of the dividend will move to the result. Thus, it is a result derived from a part (*a partibus*) – e.g., 8[00]: 36 → 8 : 4 → one fourth of 8 is sought → the result will be 2 in the order of tens, i.e., 2[0].[133]

Should we modify the dividend during division (see rules 4–9) by lowering the highest numeral of the dividend to 1, by moving it to lower orders, or by detaching a part from the dividend for subtracting lower parts of the divisor in the process of division without a difference, then in order to correctly determine the number of divisors in the given dividend, we must multiply (*multiplicare*) the gained quotient from the division of the reduced dividend by the original value of the given dividend (e.g., if we divide number 7[00] by the number 48, we decrease the value of the dividend to 1[0] and we will look for a fifth of ten, i.e., 10 1/5 = 2, by which we only gain the reduced result which must be multiplied by the original value of the dividend, i.e., $7 \cdot 2 = 14$, by which we learn that the divisor 48 is contained 14 times in the dividend 700).[134]

At the very end, Gerbert specifies to which columns results of division by tens, hundreds, and thousands are moved (*mittere*), according to the location of the dividend.[135]

Thus, Gerbert concludes his brief rules of division in which he insists that an arithmetician should minimalize actual instances of division during the

---

132 Ibid. II, 6 [10], p. 21.
133 Ibid.
134 Ibid. pp. 21–22.
135 Ibid. p. 22.

computation. These are mostly converted either to subtraction or multiplication. A division itself is limited to detaching parts from the highest numeral of the dividend (searching for halves, thirds, etc.), which can be easily transformed to multiplication by fractions. It may seem at first glance that some procedures are overly complex; however, it must be noted that following and internalising these steps can safely and accurately lead us towards correct results.

CONCLUSION

Given the fact that a substantial part of this book relies on an analysis of Gerbert's letters in which he explores various issues related to the contemporary understanding of arithmetic (even Gerbert's *Regulae de numerorum abaci* were attached to a letter to Constantine of Fleury and Micy), it would be appropriate to open the conclusion by referring to Gerbert's correspondence.

In his letter addressed to abbot Raymond (Gerbert's first teacher) and the brothers of the Aurrilac Abbey of Saint Gerald, which was written in the first half of the 990s, he includes the famous statement that the success of the disciple is the glory of the teacher (*Discipuli victoria magistri est gloria.*).[1] Gerbert adds that Raymond is the teacher to whom he is most indebted from all mortals (*omnes mortales*) for his knowledge (*scientia*).[2] Thus, Gerbert pays tribute to his teacher, although his remark about the teacher's glory can be very well applied to himself, since his own renown was intertwined with the success of his numerous students, with whom he kept up a frequent correspondence, as evidenced by the vast number of preserved letters, including those we regard as scientific.

Furthermore, a decade earlier, in his letter to abbot Evrard of Tours, Gerbert summarises the character of his own interests in knowledge and philosophy, including their importance. He rejects the attitude of the stoic philosopher Panatios of Rhodes, who sometimes (*interdum*) differentiated between the honest (*honestus*) and the useful (*utilis*). Gerbert agrees with Cicero that they should be inseparable.[3] While Panatios doubts the essential usefulness of natural philosophy, astrology, etc. (i.e., general and theoretical knowledge), Gerbert sees things differently with regard to usefulness of theoretical and practical philosophy.[4] In Gerbert's view, we should not separate (*non separare*) the theoretical (*ratio dicendi*) from the practical (*ratio morum*) in philosophy; thus, he tried

---

1   Gerbert, *Ep.* 194, p. 236. This emphasis on the significance of pedagogical practice was widespread in the Middle Ages and existed in various versions – see, e.g., Abaelard, *Ad Astral.* 907, p. 136; Bernard, *Ep.* 385, p. 351 etc. Cf. also Oldoni, M., *Gloria magistri. Orizzonti letterari e orali della cultura mediolatina.* Spoletto: Fondazione Centro italiano di studi sull'alto Medioevo, 2018.
2   Gerbert, *Ep.* 194, p. 236.
3   Idem, *Ep.* 44, pp. 72–73. Cf. p. 82 above.
4   Cf., for example, Roskam, G., *On the Path to Virtue. The Stoic Doctrine of Moral Progress and its Reception in (Middle-)Platonism.* Leuven: Leuven University Press, 2005, pp. 33–45.

instead to connect (*coniungere*) them and to strive for a good life (*studium bene vivendi*), dedicating himself to the pursuit of the true interpretation (*studium bene dicendi*). One of these might sound nobler (*praestantior*) than the other, but Gerber believed that both were necessary (*necessarius*); which is why he decided to create a library (*bibliotheca*) for these purposes in Reims.[5]

If we recall Gerbert's well-evidenced emphasis on the study of ancient authoritative sources (as clearly indicated, for example, with regard to arithmetic in both "abacist" letters to Constantine[6]) it seems clear why Gerbert's fame (*fama*)[7] spread as a result of his pedagogical activities, which were based on several fundamental pillars: a reading of elementary texts regarding all the liberal arts; an effort to maintain maximal clarity and practical utility of theoretical knowledge; the creation of illustrative, explanatory, and practically applied tools and aids; all of which was embodied by a learned man who underwent a significantly non-standard educational process that shaped his own pedagogical activities. His letter to Evrard concludes aptly – regardless of whether we have time on our hands (*otium*) or we are currently very busy (*negotium*), we should always strive to teach (*docere*) what we know (*scire*) before craving to learn (*addiscere*) what we do not know (*nescire*).[8]

The renown of Gerbert's Reims school was one of the reasons behind his stay in Ravenna, where he was invited by the emperor Otto II, at the instigation of the teacher Ohtric of Magdeburg, at the beginning of the year 981. There, Gerbert was to have defended himself in the course of a scholarly dispute from the accusation that, in his division of philosophical knowledge, he placed mathematics and physics on the same level (or he even considered mathematics to be superior to physics), whereas mathematics should be subordinate to physics,[9] according to the Stoic division of philosophy.[10] However, Gerbert preferred to follow the aforementioned Aristotelic-Boethian division of philosophy,[11] in which theoretical and practical sciences constitute the primary distinction, with the first group containing both mathematics and physics.[12] Even in this unprecedented disputation, which was attended by (according to Richer's record) numerous scholars

---

5   Gerbert, *Ep.* 44, p. 7. Cf. Lake, "Gerbert of Aurillac…," p. 63.
6   Gerbert, *Fr. de abaci*, pp. 23–24; or *Reguale*, praef., pp. 6–7.
7   Richer, *Hist.* III, 56, p. 198.
8   Gerbert, *Ep.* 44, p. 73.
9   Richer, *Hist.* III, 56, p. 199; or III, 61, p. 202.
10   Cf. Diogenes, *Vitae* VII, 39–40, p. 469; Cicero, *Tusc. disp.* V, 24, 68, p. 435; Augustine, *Civ. Dei* VIII, 10; or Isidore, *Etym.* II, 24, etc.
11   See pp. 18–19 above.
12   Richer, *Hist.* III, 60, p. 202.

of the era,[13] Gerbert proved his erudition. Apparently, it was no coincidence that Othric chose mathematics as the main subject of his charge, since Gerbert – undoubtedly inspired, among others, by Boethius's *Introduction to Arithmetic* – regarded the first of mathematical sciences, i.e., arithmetic, in particular, to be the entry point to all knowledge.

Arithmetic (i.e., knowledge of numbers), as the propaedeutic to philosophical knowledge, was one of the most important scientific disciplines for Gerbert, and while it had undoubted practical applications, it primarily led to an understanding of philosophical (metaphysical) structures of reality. It tied into Pythagorean and Platonic traditions, while it developed and rethought the Boethian basis of the arithmetical art as an essential underpinning, without which a philosopher could never become a true philosopher. Furthermore, Gerbert actively strove to learn new methods of conducting the arithmetical operations he encountered during his Catalonian studies. For future generations of medieval Latin mathematicians, he became an inspiration due to these efforts. It is worth mentioning that the procedure Gerbert recommended for adding, subtracting, etc., and the numerals he employed still occupy an uncontested place in mathematics today.

As this book has tried to demonstrate, in the eyes of Gerbert himself, arithmetic (both in its theoretical form as the philosophy of numbers, and in its practical form as abacist computing) does not represent a self-serving tool. First and foremost, the philosophy of numbers provides us with a way by which we can understand the Act of Divine Creation, while numbers (as also claimed by certain ancient philosophers) represent a metaphysical entity, the understanding of which helps us to uncover the fundamental truths of the universe in its uniqueness, and as part of a higher order. The same applies to practical arithmetic, which finds its place first and foremost among intellectual activities, since it sharpens our reasoning abilities, enables abstract thinking, and is useful for a quantitative recording of that which is discovered by the other sciences which medieval thinkers understood to be mathematical in nature – i.e., music, geometry, and astronomy, which are not only existentially dependent on arithmetic, but which also further help seek the arithmetical and metaphysical order of reality.

Theory and practice should be intertwined in perfect coexistence. Thus, it is necessary to connect (and not only in pedagogical practice) the noble with the useful, the metaphysical and theological with the ethical, the abstract with the concrete, and the speculative with the empiric. Arithmetic, as understood by Gerbert, is perfect proof of this.

---

13   Ibid. III, 57, p. 200.

BIBLIOGRAPHY

Abbreviations

References to ancient and medieval works consists of author's name and abbreviated title. All abbreviations of these works see *Primary sources* (*editions cited*) and *Index locorum*.

Series abbreviations:

    CC   *Corpus Christianorum*
          - CM   - *Continuatio Mediaevalis*
          - SL   - *Series Latina*
    CCM *Corpus consuetudinum monasticarum*
    CSEL *Corpus Scriptorum Ecclesiasticorum Latinorum*
    LCL *Loeb Classical Library*
    MGH *Monumenta Germaniae Historica*
          - Auct. ant. - *Auctores antiquissimi*
          - BDK   - *Briefe der deutschen Kaiserzeit*
          - Poetae   - *Poetae Latini medii aevi*
          - SS   - *Scriptores (in Folio)*
          - SS RG   - *Scriptores rerum Germanicarum*
          - SS RG NS - *Scriptores rerum Germanicarum, Nova series*
    PL   *Patrologiae Latinae cursus completus*

Primary Sources

Manuscripts cited

Avranches, Bibliothèque Municipale, Ms. 235
Bern, Burgerbibliothek, Cod. 250
Berlin, Staatsbibliothek, Ms. lat. oct. 162
El Escorial, Real Biblioteca del Monasterio de San Lorenzo, Cod. d.I.2
Erlangen, Universitätsbibliothek, Cod. 379
London, British Library:
        Add MS 17808
        Royal MS 13 A XI

Luxembourg, Bibliothèque nationale de Luxembourg, Ms 770
Montpellier, Bibliothèque Universitaire Historique de Médecine, H.491
Munich, Bayerische Staatsbibliothek:
    Clm MS 14436
    Clm MS 14689
Oxford, St. Johns's College, MS 17
Paris, Bibliothèque nationale de France:
    Lat. 6620
    Lat. 7189A
    Lat. 7231
    Lat. 8663
Rouen, Bibliothèque municipale, Ms. 489
St. Gallen, Stiftsbibliothek, Cod. Sang. 18
Trier, Stadtbibliothek, Hs. 1093/1694
Vatican, Biblioteca Apostolica Vaticana:
    Reg. Lat. 1661
    Vat. Lat. 644
    Vat. Lat. 3123
    Vat. Lat. 3896
    Vat. Lat. 4539

Editions cited

Abbo of Fleury (Abbo Floriacensis), *Abacus*. Ed. N. Bubnov. In: *Gerberti postea Silvestrii II papae Opera Mathematica (972–1003)*. Ed. N. Bubnov. Berlin: R. Friedländer & Sohn, 1899 [reprint Hildesheim: G. Olms, 1963], pp. 203–204.

—, *De quinque circulus mundi*. Ed. R. B. Thomson. In: Thomson, R. B., "Further Astronomical Material of Abbo of Fleury." *Mediaeval Studies* 50 (1988), pp. 671–673 [= *De circ.*].

—, *De ratione spere*. Ed. R. B. Thomson. In: Thomson, R. B., "Two Astronomical Tractates of Abbo of Fleury." In: North, J. D. – Roche, J. J. (eds.), *The Light of Nature. Essays in the History and Philosophy of Science presented to A. C. Crombie*. Dordrecht – Boston – Lancaster: Martinus Nijhoff Publishers, 1985, pp. 120–133 [= *De spere*].

—, *Explanatio in Calculo Victorii*. Ed. A. M. Peden. In: Abbo of Fleury and Ramsey, *Commentary on the Calculus of Victorius of Aquitaine*. Ed. A. M.

Peden. Oxford: Oxford University Press – The British Academy, 2003, pp. 63–136 [= *In Calc.*].

— *Excerpta in calculum Victorii commentraio*. Ed. N. Bubnov. In: *Gerberti postea Silvestrii II papae Opera Mathematica (972–1003)*. Ed. N. Bubnov. Berlin: R. Friedländer & Sohn, 1899 [reprint Hildesheim: G. Olms, 1963], pp. 197–203 [= *Excerpta*].

—, *Quaestiones grammaticales*. Ed. A. Guerrueau-Jalabert. Paris: Les Belles Lettres, 1982 [= *Quaest. gram.*].

Adelard of Bath (Adelardus Bathoniensis), *Regule abaci*. Ed. B. Boncompagni. *Bullettino di bibliografia e di storia delle scienze matematiche e fisiche* 14 (1881), pp. 91–134 [= *Reg. abaci*].

Ademar of Chabannes (Ademarus Cabannensis), *Chronicon*. Ed. J. Chavanon. Paris: A. Picard, 1897 [= *Chron.*].

Annales regni Francorum. Ed. F. Kurze. MGH SS RG 6. Hannover: Hahn, 1895 [= *Ann. Franc.*].

Aristotle (Aristotelés), Ἀθηναίων πολιτεία / *Atheniensium respublica*. Ed. M. Chambers. Stuttgart – Leipzig: Teubner, 1994 [= *Ath. pol.*].

—, *Categoriae*. In: *Aristotelis opera*. Ed. I. Bekker. Vol. I., Berlin: G. Reimer, 1831 [reed. O. Gigon. Berlin: De Gruyter, 1960], pp. 1–15 [= *Cat.*].

—, *De anima*. In: *Aristotelis opera*. Ed. I. Bekker. Vol. I., Berlin: G. Reimer, 1831 [reed. O. Gigon. Berlin: De Gruyter, 1960], pp. 402–435 [= *De an.*].

—, *Metaphysica*. In: *Aristotelis opera*. Ed. I. Bekker. Vol. I., Berlin: G. Reimer, 1831 [reed. O. Gigon. Berlin: De Gruyter, 1960], pp. 980–1093 [= *Met.*].

—, *Physica*. In: *Aristotelis opera*. Ed. I. Bekker. Vol. I., Berlin: G. Reimer, 1831 [reed. O. Gigon. Berlin: De Gruyter, 1960], pp. 184–267 [= *Phys.*].

—, *Topica*. In: *Aristotelis opera*. Ed. I. Bekker. Vol. I., Berlin: G. Reimer, 1831 [reed. O. Gigon. Berlin: De Gruyter, 1960], pp. 100–164 [= *Top.*].

Ascelin of Augsburg (Ascelinus Teutonicus), *Compositio astrolabii*. Ed. C. Burnett. In: C. Burnett, "King Ptolemy and Alchandreus the Philosopher." *Annals of Science* 55 (1998), pp. 345–351 [= *Comp. astrol.*].

Asilo of Würzburg (Asilo Wirzburgensis), *Regula de rithmachia*. Ed. A. Borst. In: Borst, A., *Das mittelalterliche Zahlenkampfspiel*. Heidelberg: C. Winter, 1986, pp. 31–34 [= *Rith.*].

Augustine (Aurelius Augustinus), *De civitate dei*. Eds. B. Dombart – A. Kalb. CCSL 48–49. Turnhout: Brepols, 1955 [= *Civ. Dei*].

—, *De Genesi ad litteram*. Ed. J. Zycha. CSEL 28. Praha – Wien – Leipzig: Tempsky – Freytag, 1894, pp. 3–456 [= *Gen. ad litt.*].

—, *De libero arbitrio*. Ed. W. M. Green. *CCSL* 29. Turnhout: Brepols, 1970, pp. 211–321 [= *Lib. arb.*].

—, *De musica*. Ed. M. Jacobsson. *CSEL* 102. Berlin – Boston: De Gruyter, 2017 [= *Mus.*].

—, *De ordine*. Ed. W. M. Green. *CCSL* 29. Turnhout: Brepols, 1970, pp. 89–137 [= *Ord.*].

Bede the Venerable (Beda Venerabilis), *De natura rerum liber*. Ed. C. W. Jones. *CCSL* 123A. Turnhout: Brepols, 1975, pp. 189–234 [= *De nat. rerum*].

—, *De temporum ratione*. Ed. C. W. Jones. *CCSL* 123B. Turnhout: Brepols, 1977 [= *De temp. rat.*].

—, *Historia ecclesiastica*. Ed. G. H. Moberly, transl. J. E. King. *LCL* 246 & 248. London – Cambridge, Mass.: W. Heinemann – Harvard University Press, 1930 [= *Hist. eccles.*].

Bernard of Clairvaux (Bernardus Claraevallensis), *Epistolae*. Ed. J. Leclercq et H.M. Rochais. In: *Sancti Bernardi Opera*. Vol. 8. Roma – Turnhout: Editiones Cistercienses – Brepols, 1977 [= *Ep.*].

Bernelin of Paris (Bernelinus iunior Parisius), *Liber abaci / Libre d'abaque. Réalisé d'après le manuscrit du XIème siècle, H.491 de la Bibliothéque de l'Ecole de Médicine de Montpellier*. Ed & trans. B. Bakhouche. Pau: Princi Néguer & C.I.H.S.O, 1999 [= *Liber abaci*].

Bibliorum Sacrorurm Editio. Nova Vulgata. Retrieved from: http://www.vatican.va/archive/bible/nova_vulgata/documents/nova-vulgata_index_lt.html.

Boethii quae dicitur geometria altera. Ed. M. Folkerts. In: Folkerts, M. *„Boethius" Geometrie II. Ein mathematisches Lehrbuch des Mittelalters*. Wiesbaden: Steiner Verlag, 1970, pp. 113–171 [= *Geom. II*].

Boethius (Anicius Manlius Torquatus Severinus), *De arithmetica*. Ed. H. Oosthout – I. Schilling. *CCSL* 94A. Turnhout: Brepols, 1999 [= *Arith.*].

—, *De consolatione philosophiae*. Ed. C. Moreschini. In: Boethius, *De consolatione philosophiae. Opuscula theologica*. München – Leipzig: K. G. Saur, ^{2}2005, pp. 3–162 [= *De cons. phil.*].

—, *De institutione musica*. Ed. G. Friedlein. Lepizig: Teubner, 1867, pp. 177–371 [= *Mus.*].

—, *De sancta trinitate*. Ed. C. Moreschini. In: Boethius, *De consolatione philosophiae. Opuscula theologica*. München – Leipzig: K. G. Saur, ^{2}2005, pp. 165–181 [= *Trin.*].

—, *In Categorias Aristotelis Commentaria*. Ed. J.-P. Migne. *PL* 64. Paris: J.-P. Migne, 1847, cols. 159–294 [= *In Cat.*].

—, *In Isagogen Porphyrii commentorum editionis primae*. Ed. G. Schepps – S-Brandt. *CSEL* 48. Wien – Leipzig: Tempsky – Freytag, 1906, pp. 3–132 [= *1 In Isag.*].

Byrhtfert of Ramsey (Bridfertus Ramesiensis), *De loquela per gestum digitorum et temporum ratione libellus*. Ed. J.-P. Migne. *PL* 90. Paris: J.-P. Migne, 1862, cols. 685–698 [= *De loquela*].

Calcidius (Chalcidius), *Commentarius in Platonis Timaeum*. Ed. J. H. Waszink. In: *Timaeus a Calcidio translatus commentarioque instructus*. Ed. J. H. Waszink. London – Leiden: The Warburg Institute – Brill, 1975, pp. 57–346 [= *In Tim.*].

Cassiodore (Flavius Magnus Aurelius Cassiodorus), *Institutiones*. Ed. R. A: B. Mynors. Oxford: Clarendon Press, ²1961 [= *Inst.*].

—, *Variae*. Ed. T. Mommsen. *MGH Auct. ant.* 12. Berlin: Weidmann, 1894 [= *Var.*].

Censorinus, *De die natali liber*. Ed. N. Sallmann. Leipzig: Teubner, 1983 [= *Die nat.*].

Ciacono, Alphons (Alphonsus Ciacconius, Chacón) – Oldoini, Augustine (Augustinus Oldoinus), *Vitae, et res gestae pontificum romanorum et S.R.E. cardinalium ab initio nascentis Ecclesiae usque ad Clementem IX. P.O.M.* Roma: J. B. Bernabo & J. Lazzarini, 1677 [= *Vitae pontif. card.*].

Cicero, Marcus Tullius, *Tusculanae disputationes*. Ed. M. Pohlenz. Berlin: De Gruyter, 2008 [= *Tusc. disp.*].

Commentarii in Gerberti regulas de numerorum abaci rationibus. Ed. N. Bubnov. In: *Gerberti postea Silvestrii II papae Opera Mathematica (972–1003)*. Ed. N. Bubnov. Berlin: R. Friedländer & Sohn, 1899 [reprint Hildesheim: G. Olms, 1963], pp. 245–284 [= *In Gerb.*].

Consuetudines Floriacenses antiquiores. Eds. A. Davril – L. Donnat. In: *Consuetudinum saeculi X/XI/XII monumenta non-Cluniacensia*. Ed. K. Hallinger. *CCM* 7, 3. Sieburg: F. Schmitt, 1984, pp. 7–60 [= *Consuet. Flor.*].

De arithmetica Boetii. Ed. I. Caiazzo. In: Caiazzo, I., "Un commento altomedievale al *De arithmetica* di Boezio." *Archivum Latinitas Medii Aevi* 58 (2000), pp. 126–150 [= *De arith. Boeth.*].

De minutiis. Ed. N. Bubnov. In: *Gerberti postea Silvestrii II papae Opera Mathematica (972–1003)*. Ed. N. Bubnov. Berlin: R. Friedländer & Sohn, 1899 [reprint Hildesheim: G. Olms, 1963], pp. 225–244.

De mundi coelestis terrestrisque constitutione liber. Ed. J.-P. Migne. *PL* 90, Paris: J.-P. Migne, 1850, cols. 881–910 [= *De mundi*].

Diogenes Laertius (Diogenés Laertios), *Vitae philosophorum*. Ed. M. Marcovich. Stuttgart – Leipzig: Teubner, 1999 [= *Vitae*].

Epaphroditus (Epafroditos) – Vitruvius Rufus (Betrubus Rufus) [?], *Excerpta*. In: *Gerberti postea Silvestrii II papae Opera Mathematica (972-1003)*. Ed. N. Bubnov. Berlin: R. Friedländer & Sohn, 1899 [reprint Hildesheim: G. Olms, 1963], pp. 516–551.

Epistolae quorundam. Ed. N. Bubnov. In: *Gerberti postea Silvestrii II papae Opera Mathematica (972-1003)*. Ed. N. Bubnov. Berlin: R. Friedländer & Sohn, 1899 [reprint Hildesheim: G. Olms, 1963], pp. 285–290 [= *Ep. quord.*].

Epitaphium Pacifici archidiaconi. Ed. E. Dümmler. In: *Poetae Latini aevi Carolini*. T. 2. MGH Poetae 2. Berlin: Weidmann, 1884, pp. 655–656 [= *Epit. Pacif.*].

Eratosthenes of Cyrene (Erathostenés Kyrénaios/Cyrenaeus), *Die geographischen Fragmente des Eratosthenes*. Ed. H. Berger. Leipzig: Teubner, 1880 [= *Fragm.*].

Eriugena, John Scotus (Iohannes Scotus Eriugena), *Annotationes in Marcianum*. Ed. C. E. Lutz. Cambridge, Mass.: The Mediaeval Academy of America, 1939 [= *In Marc.*].

—, *Periphyseon (De divisione naturae)*. 5 Vols. Ed. É. Jeauneau. CCCM 161–165. Turnhout: Brepols, 1996–2003 [*De div.*].

Euclid (Eukleidés). *Euclidis Elementa*. Eds. I. L. Heiberg – E. S. Stamatis. Vol. 1–5. Leipzig: De Gruyter, 1969–1977 [= *Elem.*].

Excerptiuncula. Ed. C. V. Crialesi. In: Crialesi, C. V., "The *Excerptiuncula*: A Short Introduction to Boethius's *De artihemtica* from the Early Middle Ages." *The Journal of Medieval Latin* 31 (2021), pp. 279–282 [= *Excerpt.*].

Geometria incerti auctoris. In: *Gerberti postea Silvestrii II papae Opera Mathematica (972-1003)*. Ed. N. Bubnov. Berlin: R. Friedländer & Sohn, 1899 [reprint Hildesheim: G. Olms, 1963], pp. 309–364 [= *Geom. inc.*].

Gerbert of Aurillac/of Reims, Pope Sylvester II (Gerbertus Auriliacensis/Remensis), *Ad Adelboldum de causa diversitatis arearum trigoni aequlateri geometrice arithmeticeve expensi*. In: *Gerberti postea Silvestrii II papae Opera Mathematica (972-1003)*. Ed. N. Bubnov. Berlin: R. Friedländer & Sohn, 1899 [reprint Hildesheim: G. Olms, 1963], pp. 41–45 [= *Ad Adel.*].

—, *De sphaera*. Ed. N. Bubnov. In: *Gerberti postea Silvestrii II papae Opera Mathematica (972-1003)*. Ed. N. Bubnov. Berlin: R. Friedländer & Sohn, 1899 [reprint Hildesheim: G. Olms, 1963], pp. 24–28.

—, *De utilitatibus astrolabii*. Ed. N. Bubnov. In: *Gerberti postea Silvestrii II papae Opera Mathematica (972-1003)*. Ed. N. Bubnov. Berlin: R. Friedländer & Sohn, 1899 [reprint Hildesheim: G. Olms, 1963], pp. 109–147 [= *De util. astrol.*].

—, *Die Briefsammlung*. Ed. F. Weigle. MGH BDK 2. Weimar: Hermann Böhlau, 1966 [= *Ep.*].

- *Correspondance*. Ed. & Transl. P. Riché – J.-P. Callu. Paris: Les Belles Lettres, 2008.
- *Epistolario*. Transl. M. G. Panvini-Carciotto – C. Sigismondi – P. Rossi. Roma: APRA, 2010.
- *Lettere (983-997)*. Transl. P. Rossi. Pisa: Plus – Pisa University Press, 2009.
- *The Letters of Gerbert with His Papal Privileges as Sylvestre II*. Trans. H. Pratt Lattin. New York: Columbia University Press, 1961.

—, *Omnigenum pater*.... Ed. G. Silagi. In: *Die Lateinischen Dichter des deutschen Mittelalters*. Ed. G. Silagi. MGH Poetae 5/3. München: MGH, 1979, pp. 666-667 [= *Figur.*].

—, *Fragmentum de norma rationis abaci*. Ed. N. Bubnov. In: *Gerberti postea Silvestrii II papae Opera Mathematica (972-1003)*. Ed. N. Bubnov. Berlin: R. Friedländer & Sohn, 1899 [reprint Hildesheim: G. Olms, 1963], pp. 23-24 [= *Fr. de abaci*].

—, *Geometria*. Ed. N. Bubnov. In: *Gerberti postea Silvestrii II papae Opera Mathematica (972-1003)*. Ed. N. Bubnov. Berlin: R. Friedländer & Sohn, 1899 [reprint Hildesheim: G. Olms, 1963], pp. 46-97 [= *Geom.*].

—, *Regulae de numerorum abaci*. Ed. N. Bubnov. In: *Gerberti postea Silvestrii II papae Opera Mathematica (972-1003)*. Ed. N. Bubnov. Berlin: R. Friedländer & Sohn, 1899 [reprint Hildesheim: G. Olms, 1963], pp. 6-22 [= *Regulae*].

- *Libellus de numerorum divisione*. Ed. A. Olleris. In: *Œuvres de Gerbert pape sous le nom de Sylvestre II*. Ed. A. Olleris. Clermont-Ferrand – Paris: F. Thibaud – Ch. Dumoulin, 1867, pp. 349-355 [= *De num. div.*].
- *Traité de Gerbert*. Ed & trans. M. Chasles. In: Chasles, M. "Analyse et explication du traité de Gerbert." *Comptes Rendus des séances de l'Académie des Sciences* 16 (1843), pp. 295-299 [= *Traité*].

—, *Rogatus a pluribus*. Ed. K.-J. Sachs. In: Sachs, K.-J., *Mensura fistularum. Die Mensurierung der Orgelpfeifen im MIttelalter*. Stuttgart: Musikwissenschaftliche Verlags-Gesellschaft, 1970, pp. 59-81 [= *Rogatus*].

—, *Scholium ad Boethii Arithmeticam Institutionem l. II, c. I*. Ed. N. Bubnov. In: *Gerberti postea Silvestrii II papae Opera Mathematica (972-1003)*. Ed. N. Bubnov. Berlin: R. Friedländer & Sohn, 1899 [reprint Hildesheim: G. Olms, 1963], pp. 31-35 [= *In Boeth. Arith.*].

—, *Scholium ad Boethii Musicae Institutionis l. II, c. 10; l. IV, c. 2*. Ed. N. Bubnov. In: *Gerberti postea Silvestrii II papae Opera Mathematica (972-1003)*. Ed. N. Bubnov. Berlin: R. Friedländer & Sohn, 1899 [reprint Hildesheim: G. Olms, 1963], pp. 28-30 [= *1 In Boeth. Mus.*].

—, *Scholium ad Boethii Musicae Institutionis l. II, c. 21*. Ed. N. Bubnov. In: *Gerberti postea Silvestrii II papae Opera Mathematica (972–1003)*. Ed. N. Bubnov. Berlin: R. Friedländer & Sohn, 1899 [reprint Hildesheim: G. Olms, 1963], pp. 30–31 [= *2 In Boeth. Mus.*].

Glosses to Boethius' De Arithmetica in Cologne Ms. 186. Ed. H. Mayr-Harting. In: H. Mayr-Harting, *Church and Cosmos in Early Ottonian Germany. The View from Cologne*. Oxford: Oxford University Press, 2007, pp. 248–270 [= *In Boeth. Arith.*].

Heriger of Lobbes (Herigerus Lobiensis), *Regulae de numerorum abaci rationibus*. Ed. N. Bubnov. In: *Gerberti postea Silvestrii II papae Opera Mathematica (972–1003)*. Ed. N. Bubnov. Berlin: R. Friedländer & Sohn, 1899 [reprint Hildesheim: G. Olms, 1963], pp. 205–221 [= *Regulae*].

- *Regula de abaco computi*. Ed. A. Olleris. In: *Œuvres de Gerbert, pape sous le nom de Sylvestre II*. Ed. A. Olleris. Clermont-Ferrand – Paris: F. Thibaud – C. Dumoulin, 1867, pp. 311–348 [= *Regula*].

Hermann of Reichenau (Hermannus Augiensis/Contracus), *De mensura astrolabii*. Ed. J. Drecker. In: J. Drecker, "Hermannus Contractus – Über das Astrolab." *Isis* 16/2 (1931), pp. 203–212 [= *De mens. astrol.*].

—, *Regula de rithmimachia*. Ed. A. Borst. In: Borst, A., *Das mittelalterliche Zahlenkampfspiel*. Heidelberg: C. Winter, 1986, pp. 35–39 [= *Rith.*].

Hyginus, Gaius Julius (astronomus), *De astronomia*. Ed. G. Viré. Stuttgart – Leipzig: Teubner, 1992 [= *De astron.*].

Iamblichus of Chalcis (Iamblichos Chalcidensis). *Iamblichi In Nicomachi Arithmeticam Introductionem liber*. Ed. H. Pistelli. Leipzig: Teubner, 1894 [= *In Nic. Arith.*].

Isidore of Sevilla (Isidorus Hispalensis), *Etymologiarum sive Originum libri XX*. Ed. W. M. Lindsay. Oxford: Clarendon Press, 1911 [= *Etym.*].

Macrobius (Ambrosius Theodosius), *Commentarii in Somnium Scipionis*. Ed. J. Willis. Stuttgart – Leipzig: Teubner, 1994 [= *In Somn.*].

Martianus Capella, *De nuptiis Philologiae et Mercurii*. Ed. J. Willis. Leipzig: Teubner, 1983 [= *De nupt.*].

Nicomachus of Gerasa (Nicomachus Gerasenus), *Introductio arithmeticae*. Ed. R. Hoche. Leipzig: Teubner, 1866 [= *Arith.*].

Notker of Liège (Notgerus Leodiensis), *De superparticulari*. Ed. N. Bubnov. In: *Gerberti postea Silvestrii II papae Opera Mathematica (972–1003)*. Ed. N. Bubnov. Berlin: R. Friedländer & Sohn, 1899 [reprint Hildesheim: G. Olms, 1963], pp. 297–299 [= *Superpart.*].

Pacificus of Verona (Pacificus Veronensis), *Spera caeli*. Ed. K. Stecker. *MGH Poetae* IV/2. Berlin: Weidmann, 1923, p. 692 [= *Spera*].

Peter Abelard (Petrus Abaelardus), *Carmen ad Astralabium*. Ed. J. F. Ruys. In: Ruys, J. F., *The Repentant Abelard: Family, Gender, and Ethics in Peter Abelard's Carmen ad Astralabium and Planctus*. New York: Palgrave Macmillan, 2014, pp. 93–142 [= *Ad Astral.*].

Plato (Platón), *Gorgias*. In: *Platonis Opera*. Ed. J. Burnett. T. 3. Oxford: Clarendon Press, 1903, pp. 447–527 [= *Gorg.*].

—, *Respublica*. T. 4. Oxford: Clarendon Press, 1904, pp. 327–621 [= *Resp.*].

—, *Theaetetus*. In: *Platonis Opera*. Ed. J. Burnett. T. 1. Oxford: Clarendon Press, 1903, pp. 142–210 [= *Tht.*].

—, *Timaeus*. In: *Platonis Opera*. Ed. J. Burnett. T. 4. Oxford: Clarendon Press, 1904, pp. 17–105 [= *Tim.*].

Pliny the Elder (Gaius Plinius Secundus), *Naturalis historiae libri XXXVII*. 5 vols. Eds. L. Jan - K. Mayhoff. Leipzig: Teubner 1892–1909 [= *Nat. hist.*].

Ptolemy (Klaudios Ptolemaios), *Syntaxis mathematica*. In: *Claudii Ptolemaei opera quae extant omnia*. Vol. 1–2. Ed. J. L. Heiberg. Leipzig: Teubner, 1898 [= *Alm.*].

Quid sit abacus. Ed. P. Treutlein. *Bullettino di bibliografia e di storia delle scienze matematiche e fisiche* 10 (1877), pp. 625–630 [= *Abacus*].

Ralph of Laon (Radulphus Laudunensis), *Liber de abaco*. Ed. A. Nagl. *Zeitschrift für Mathematik und Physik* 34/supp. 5 (1890), pp. 96–133 [= *De abaco*].

Regulae arthimeticae. Ed. P. Treutlein. *Bullettino di bibliografia e di storia delle scienze matematiche e fisiche* 10 (1877), pp. 607–625 [= *Reg. arith.*].

Regulae domni Odonis super abacum. Ed. J.-P. Migne. *PL* 133. Paris: Gernier, 1853, cols. 807–814 [= *Odon. super abacum*].

Remigius of Auxerre (Remigius Autissiodorensis), *Commentum in Martianum Capellam*. Ed. C. E. Lutz. 2 Vols. Leiden: Brill, 1962, 1965 [= *In Marc.*].

Richer of Reims (Richerus Remensis), *Historiarum libri IIII*. Ed. H. Hoffmann, *MGH SS* 38. Hannover: Hahn, 2000 [= *Hist.*].

Stobaeus, John (Ióannés Stobaios), *Eclogae physicae et ethicae*. Ed. C. Wachsmuth. Berlin: Weidmann, 1884 [= *Ecl.*].

Thietmar of Merseburg (Thietmarus Merseburgensis), *Chronicon*. Ed. R. Holtzmann. *MGH SS RG NS*. T. 9. Berlin: Weidmann, 1935 [= *Chron.*].

Tractatus de divisione. Ed. P. Treutlein. *Bullettino di bibliografia e di storia delle scienze matematiche e fisiche* 10 (1877), pp. 630–639 [= *Tract. de div.*].

Turchill Compotista (Turchillus Compotista), *Reguncule super abacum*. Ed. E. Narducci. *Bullettino di bibliografia e di storia delle scienze matematiche e fisiche* 15 (1882), pp. 135–154 [= *Reg. super abacum*].

Victorius of Aquitaine (Victorius Aquitanus), *Calculus*. Ed. A. M. Peden. In: Abbo of Fleury and Ramsey, *Commentary on the Calculus of Victorius of Aquitaine*. Ed. A. M. Peden. Oxford: Oxford University Press – The British Academy, 2003, pp. 1–62 [= *Calc.*].

William of Malmesbury (Willelmus Malmesbiriensis), *Gesta regum Anglorum / The History of English Kings*. Ed. & trans. R. A. B. Mynors. Vol. I–II. Oxford: Clarendon Press 1998 [= *Gesta reg.*].

Wyon, Arnold (Arnoldus Wionus), *Lignum vitae, ornamentum et decus ecclesiae*. Venice: G. Angelerius, 1595 [= *Lignum vitae*].

Secondary Sources

Albertson, D., "*Boethius Noster:* Thierry of Chartres's *Arithmetica* Commentary as a Missing Source of Nicholas of Cusa's *De docta ignorantia*." *Recherches de Théologie et Philosophie médiévales* 83/1 (2016), pp. 143–199.

Ambrosetti, N., *L'eredità arabo-islamica nelle scienze e nelle arti del calcolo dell'europa medievale*. Milano: Edizioni Universitarie di Lettere Economia Diritto, 2008.

Ampère, J.-J., *Histoire littéraire de la France avant le douzième siècle*. Vol. 3. Paris: Hachette, 1840.

—, "Histoire littéraire de la France avant le douzième siècle." *Revue des Deux Mondes* 5 (1836), pp. 28–38.

Bai, J., "The Spectrum of the Divine Order: Goodness, Beauty, and Harmony." *Soundings: An Interdisciplinary Journal* 102/1 (2019), pp. 1–30.

—, "Numbers: Harmonic Ratios and Beauty in Augustinian musical Cosmology." *Cosmos and History: The Journal of Natural and Social Philosophy* 13/3 (2017), pp. 192–217.

Bakhouche, B., "Introduction." In: Bernelin, élève de Gerbert d'Aurillac, *Libre d'Abaque*. Pau: Princi Néguer, 1999, pp. 9–13.

Barker, A., "Music and Perception: A Study in Aristoxenus." *The Journal of Hellenic Studies* 98 (1978), pp. 9–16.

Barnes, J., "Boethius and the Study of Logic." In: Gibson, M. (ed.), *Boethius. His Life, Thought and Influence*. Oxford: Basil Blackwell, 1981, pp. 73–89.

Baumann, H., *Die Ottonen*. Stuttgart: Kohlhammer, 1987.

Beaujouan, G., "Les Apocryphes mathématiques de Gerbert." In: Tosi, M. (ed.), *Gerberto – scienza, storia e mito. Atti del Gerberti Symposium*. Bobbio: A.S.B., 1985, pp. 645–658.

Benson, R. L. – Constable, G. (eds.), *Renaissance and renewal in the twelfth cenutry*. Toronto – Buffalo – London: Univeristy of Toronto Press, 1999.

Bergmann, W., "Gerbert von Aurillac und die Landvermessung." In: Junius, H. (ed.), *Ingenieurvermessung von der Antike bis zur Neuzeit*. Stuttgart: Wittwer, 1987, pp. 108–140.

—, *Innovationen im Quadrivum des 10. und 11. Jahrhunderts. Studien zur Einführung von Astrolab und Abakus im Lateinischen Mittelalter*. Stuttgart: Steiner Verlag, 1985.

Berschin, W. – Hellmann, M., *Hermann der Lahme. Gelehrter und Dichter (1013-1054)*. Heidelberg: Mattes, ²2005.

Borst, A., *Computus: Zeit und Zahl in der Geschichte Europas*. Berlin: Wagenbach, 1990.

—, *Astrolab und Klosterreform an der Jahrtausendwende*. Heidelberg: C. Winter, 1989.

—, *Das mittelalterliche Zahlenkampfspiel*. Heidelberg: C. Winter, 1986.

—, "Ein Forschungsbericht Hermanns des Lahmen." *Deutsches Archiv für Erforschung des Mittelalters* 40 (1984), pp. 379–477.

Bower, C. M., "The Modes of Boethius." *The Journal of Musicology* 3/3 (1984), pp. 252–263.

—, "Boethius and Nichomachus: An Essay Concerning the Sources of *De institutione musica*." *Vivarium* 16/1 (1978), pp. 1–45.

Brockett, W., "The Frontispiece of Paris, Bibliothèque Nationale Ms. Lat. 776. Gerbert's Acrostic Pattern Poems." *Manuscripta* 39 (1995), pp. 3–25.

Brown, N. M., *The Abacus and the Cross. The Story of the Pope Who Brought the Light of Science to the Dark Ages*. New York: Basic Books, 2010.

Burkert, W., *Weisheit und Wissenschaft. Studien zu Pythagoras, Philolaos und Platon*. Nürnberg: Hans Carl, 1962.

Burnett, C., "The Abacus at Echternach in ca. 1000 A.D." *SCIAMVS* 3 (2002), pp. 91–108.

—, "King Ptolemy and Alchandreus the Philosopher: The Earliest Texts on the Astrolabe and the Arabic Astrology at Fleury, Micy and Chartres." *Annals of Science* 55 (1998), pp. 329–368.

Burnett, C. – Ryan, W. F., "Abacus (Western)." In: Bud, R. – Warner, D. J. (eds.), *Instruments of Sciences. An Historical Encyclopedia*. New York – London: Garland Publishing et al., 1998, pp. 5–7.

Caiazzo, I., "Medieval Commentaries on Boethius's *De arithmetica*: A Provisional Handlist." *Bulletin de philosophie médiévale* 62 (2020), pp. 3–13.

—, "Un commento altomedievale al *De arithmetica* di Boezio." *Archivum Latinitatis Medii Aevi* 58 (2000), pp. 113–150.

Caldwell, J., "The *De Institutione Arithmetica* and the *De Institutione Musica*." In: Gibson, M. (ed.), *Boethius. His Life, Thought and Influence*. Oxford: Basil Blackwell, 1981, pp. 135–155.

Cantor, M., *Die römischen Agrimensoren und ihre Stellung in der Geschichte der Feldmesskunst. Eine historisch-mathematische Untersuchung*. Leipzig: Teubner, 1875.

Carozzi, C., "Gerbert et le concile de St-Basle." In: Tosi, M. (ed.), *Gerberto – scienza, storia e mito. Atti del Gerberti Symposium*. Bobbio: A.S.B., 1985, pp. 661–676.

Catalani, L., "«Omnia Numerorum Videntur Ratione Formata». A 'Computable World' Theory in Early Medieval Philosophy." In: Gadducci, F. – Tavosanis, M. (eds.), *History and Philosophy of Computing*. Berlin – Cham: Springer, 2016, pp. 131–140.

Celhoffer, M., "Hudba, animus a proporce v kontextu Marchettova Lucidaria aneb „svůdné vábení Sirén"." *Musicologica Brunensia* 46/1-2 (2011), pp. 49–54.

Chadwick, H., *Boethius. The Consolations of Music, Logic, Theology, and Philosophy*. Oxford: Clarendon Press, 1981.

Chasles, M., "Analyse et explication du traité de Gerbert." *Comptes Rendus des séances de l'Académie des Sciences* 16 (1843), pp. 281–299.

Cherniss, H., *Aristotle's Criticism of Presocratic Philosophy*. New York: Octagon Books, 1971.

Contreni, J., "The Caroligian Renaissance: Education and Literary Culture." In: McKitterick, R. (ed.), *The New Cambridge Medieval History*. Vol. 2: *c. 700– c. 900*. Cambridge: Cambridge University Press, pp. 709–757.

Cornelli, G., *In Search of Pythagoreanism. Pythagoreanism as an Historiographical Category*. Berlin – Boston: De Gruyter, 2013, pp. 137–187.

Corti, L., "Scepticism, number and appearances. The ἀριθμητικὴ τέχνη and Sextus' targets in M I-VI." *Philosophie antique* 15 (2015), pp. 121–145.

Crialesi, C. V., "The *Excerptiuncula*: A Short Introduction to Boethius's *De artihemtica* from the Early Middle Ages." *The Journal of Medieval Latin* 31 (2021), pp. 265–287.

—, "The Status of Mathematics in Boethius: Remarks in the Light of his Commentaries on the *Isagoge*." In: Giovannetti, L. (ed.), *The Sustainability of Thought: An Itinerary through the History of Philosophy*. Napoli: Bibliopolis, 2020, pp. 95–124.

Crossley, J. N., "The Writings of Boethius and the Cogitations of Jacobus de Ispania on Musical Proportions." *Early Music History* 36 (2017), pp. 14–24.

D'Onofrio, G., "Introduzione." In: D'Onofrio, G. (ed.), *Excerpta Isagogarum et Categoriarum*. CCCM 120. Turnhout: Brepols, 1995, pp. VII–CXVIII.

Dachowski, E., *First Among Abbots: The Career of Abbo of Fleury.* Washington: Catholic University of America Press, 2008.

Darlington, O. G., "Gerbert, the Teacher." *The American Historical Review* 52/3 (1947), pp. 456–476.

De Rijk, L. M., "On the chronology of Boethius' works on logic I." *Vivarium* 2 (1964), pp. 1–49.

—, "On the chronology of Boethius' works on logic II." *Vivarium* 2 (1964), pp. 125–161.

Dekker, E., *Illustrating the Phaenomena. Celestial Cartography in Antiquity and the Middle Ages.* Oxford: Oxford University Press, 2013.

Delville, J.-P. – Kupper, J.-L.– Laffineur-Crepin, M. (eds.), *Notger et Liège. L'an mil au cœur de l'Europe.* Liège: Éditions du Perron, 2008.

DeMayo, C., "The Students of Gerbert of Aurillac's Cathedral School at Reims: An Intellectual Genealogy." *Medieval Prosopography* 27 (2012), pp. 97–117.

Deza, E. – Deza, M. M., *Figurate Numbers.* Singapore – Hackensack: World Scientific, 2012.

Dohrn-van Rossum, G., *History of the Hour: Clocks and Modern Temporal Orders.* Chicago – London: University of Chicago Press, 1996.

Eastwood, B. S., "Astronomy in Christian Latin Europe c. 500–c. 1150." *Journal for the History of Astronomy* 28 (1997), pp. 235–258.

Engelen, E.-M., *Zeit, Zahl und Bild. Studien zur Verbindung von Philosophie und Wissenschaft bei Abbo von Fleury.* Berlin – New York: De Gruyter, 1993.

Erlande-Brandenburg, A. – Cazeaux, M. (eds.), *Le temps de Fulbert; Fulbert et les écoles de Chartres. Actes de l'Université d'été du 8 au 10 juillet 1996.* Chartres: Société Archéologique d'Eure-et-Loir, 1996.

Evans, G. R., "Boethius' Geometry and the Four Ways." *Centaurus* 25/2 (1981), pp. 161–165.

—, "The *Saltus Gerberti*: The Problem of the 'Leap'." *Janus* 67 (1980), pp. 261–268.

—, "Schools and scholars: the study of the abacus in English Schools c. 980–c. 1150." *The English Historical Review* 94 (1979), pp. 71–89.

—, "Introductions to Boethius's "Arithmetica" of the Tenth to the Fourteenth Century." *History of Science* 16 (1978), pp. 22–41.

—, "Difficillima et Ardua: theory and practice in treatises on the abacus, 950–1150." *Journal of Medieval History* 3 (1977), pp. 21–38.

Evans, G. R. – Peden, A. M., "Natural Science and Liberal Arts in Abbo of Fleury's Commentary on the Calculus of Victorius of Aquitaine." *Viator* 16 (1985), pp. 109–127.

Ferguson, W. K., *The Renaissance in Historical Thought. Five Centuries of Interpretation.* Boston – New York et al.: Houghton Mifflin Company, 1948.

Flusche, A. M., *The Life and Legend of Gerbert of Aurillac. The Organbuilder Who Become Pope Sylvester II.* Lewiston: Edwin Mellen Press, 2005.

Folkerts, M., "Die Mathematik der Agrimensoren – Quellen und Nachwirkung." In: Möller, C. – Knobloch, E., (eds.), *In den Gefilden der römischen Feldmesser. Juristische, wissenschaftsgeschichtliche, historische und sprachliche Aspekte.* Berlin – Boston: De Gruyter, 2013, pp. 131–148.

—, "The *Geometry II* Ascribed to Boethius." In: idem, *Essays on Early Medieval Mathematics. The Latin Tradition.* Aldershot: Ashgate, 2003, pp. IX-1–9.

—, "'Rithmomachia', a Mathematical Game from the Middle Ages." In: idem, *Essays on Early Medieval Mathematics. The Latin Tradition.* Aldershot: Ashgate, 2003, pp. XI-1–23.

—, "Early Texts on Hindu-Arabic Calculation." *Science in Context* 14/1–2 (2001), pp. 13–38.

—, "The names and forms of the numerals on the abacus in the Gerbert tradition." In: Nuvolone, F. G. (ed.), *Gerberto d'Aurillac da Abate di Bobbio a Papa dell'Anno 1000. Atti del Congresso internazionale.* Bobbio: Archivum Bobiense, 2001, pp. 245–265.

—, "Frühe Darstellungen des Gerbertschen Abakus." In: Franci, R. – Pagli, P. – Rigatelli, L. T. (eds.), *Itinera mathematica. Studi in onore di Gino Arrighi per il suo 90° compleanno.* Sienna: Univesità di Siena, 1996, pp. 23–43.

—, "Mathematische Probleme im Corpus agrimensorum." In: Behrends, O. – Campogrossi Colognesi, L. (eds.), *Die römische Feldmeßkunst. Interdisziplinäre Beiträge zu ihrer Bedeutung für die Zivilisationsgeschichte Roms.* Göttingen: Vandenhoeck Ruprecht, 1992, pp. 311–334.

—, *„Boethius" Geometrie II. Ein mathematisches Lehrbuch des Mittelalters.* Wiesbaden: Steiner Verlag, 1970.

Folkerts, M. – Hughes, B., "The Latin Mathematics of Medieval Europe." In: Katz, V. J. (ed.), *Sourcebook in the Mathematics of Medieval Europe and North Africa.* Princeton: Princeton University Press, 2016, pp. 4–223.

Fournier, M., "Boethius and the Consolation of the Quadrivium." *Medievalia et Humanistica. Studies in Medieval and Renaissance Culture* 34 (2008), pp. 1–21.

Franci, R., "L'insegnamento dell'aritmetica nel Medioevo." In: Franci, R. – Pagli, P. – Rigatelli, L. T. (eds.), *Itinera mathematica. Studi in onore di Gino Arrighi per il suo 90° compleanno.* Siena: Univeisità di Siena, 1996, pp. 111–132.

Frassetto, M. (ed.), *The Year 1000: Religious nad Social Response to the Turning of the First Millennium.* New York: Palgrave Macmillan, 2002.

Friedlein, G., *Die Zahlzeichen und das elementare Rechnen der Griechen un Römer und des christlichen Abendlandes vom 7. bis 13. Jahrhundert.* Erlangen: A. Deichert, 1869.

—, "Gerbert's Regeln der Division." *Zeitschrift für Mathematik und Physik* 9 (1864), pp. 145–166.

Frova, C., "Gerberto philosophus: il De rationali et ratione uti." In: Tosi, M. (ed.), *Gerberto – scienza, storia e mito. Atti del Gerberti Symposium.* Bobbio: A.S.B., 1985, pp. 351–377.

Genin, C., *Fulbert de Chartres (vers 970–1028): Une grande figure de l'Occident chrétien au temps de l'an Mil.* Chartres: Société Archéologique d'Eure-et-Loir, 2003.

Germann, N., *De Temporum Ratione. Quadrivium und Gotteserkenntnis am Beispiel Abbos von Fleury und Hermanns von Reichenau.* Leiden: Brill, 2006.

Gibson, S., *Aristoxenus of Tarentum and The Birth of Musicology.* New York – London: Routledge, 2005.

Glenn, J., *Politics and History in the Tenth Century. The Work and World of Richer of Reims.* Cambridge: Cambridge University Press, 2004.

Goold, G. P., "Introduction." In: Manilius, M., *Astronomica.* Ed. and trans. G. P. Goold. *LCL* 469. Cambridge: Harvard University Press, 1977, pp. xi–cxxiii.

Guillaumin, J.-Y., "Boethius's De Institutione Arithmetica and Its Influence on Posterity." In: Kaylor, N. H. Jr. – Phillips, P. E. (eds.), *A Companion to Boethius in the Middle Ages.* Leiden – Boston: Brill, 2012, pp. 135–161.

—, "Les deux définitions de l'angle plan par Gerbert." In: Callebat, L. – Desbordes, O. (eds.), *Science antique - science médiévale (autour d'Avranches 235).* Hildesheim: Olms-Weidmann, 2000, pp. 359–372.

Gurd, S. A., *The Origins of Music Theory in the Age of Plato.* London – New York: Bloomsbury Academic, 2019.

Harrison, C., *On Music, Sense, Affect and Voice.* London – New York: Bloomsbury Academic, 2019.

—, "Augustine and the art of music." In: Begbie, J. S. – Guthrie, S. R. (eds.), *Resonant witness. Conversations between Music and Theology.* Grand Rapids – Cambridge: William B. Eerdmans, 2001, pp. 27–45.

Haskins, Ch. H., *The Renaissance of the Twelfth Century.* Cambridge: Harvard University Press, 1927.

Head, T., "Letaldus of Micy and the Hagiographic Traditions of the Abbey of Nouaillé. The Context of the *Delatio corporis S. Juniani.*" *Analecta Bollandiana* 115/3-4 (1997), pp. 253–267.

Heinzer, F. – Zotz, T. (eds.), *Hermann der Lahme. Reichenauer Mönch und Universalgelehrter des 11. Jahrhunderts*. Stuttgart: Kohlhammer, 2016.

Henderson, J., *The Medieval World of Isidore of Seville. Truth from Words*. Cambridge: Cambridge University Press, 2007.

Herlinger, J. W., *The Lucidarium of Marchetto of Padua. A Critical Edition, Translation, and Commentary*. Chicago: University of Chicago, 1985.

Herren, M., "Classical and Secular Learning among the Irish before the Carolingian Renaissance." *Florilegium* 3 (1981), pp. 118–157.

Hiatt, A., "The Map of Macrobius before 1100." *Imago Mundi* 59/2 (2007), pp. 149–176.

Hicks, A., *Composing the World. Harmony in the Medieval Platonic Cosmos*. Oxford: Oxford University Press, 2017.

—, "Martianus Capella and the Liberal Arts." In: Hexter, R. – Townsend, D. (eds.), *The Oxford Handbook of Medieval Latin Literature*. Oxford: Oxford University Press, 2012, pp. 307–334.

Hill, D. R., "Clocks and Watches." In: Selin, H. (ed.), *Encyclopaedia of the History of Science, Technology, and Medicine in Non-Western Cultures*, Volume 1: A–K. Berlin – New York: Springer, 2008, pp. 151–153.

Holmes, U. T., "The Idea of a Twelfth-Century Renaissance." *Speculum* 26/4 (1951), pp. 643–651.

Honigman, E., *Die sieben Klimata und die ΠΟΛΕΙΣ ΕΠΙΣΗΜΟΙ*. Heidelberg: C. Winter. 1929.

Høyrup, J., "Mathematics Education in the European Middle Ages." In: Karp, A. – Schubring, G. (eds.), *Handbook on the History of Mathematics Educations*. New York: Springer, 2014, pp. 109–124.

Huffman, C. A. (ed.), *Aristoxenus of Tarentum. Discussion*. New Brunswick: Transaction Publishers, 2012.

—, "Two Problems in Pythagoreanism." In: Curd, P. – Graham, D. W. (eds.), *The Oxford Handbook of Presocratic Philosophy*. Oxford: Oxford University Press, 2008, pp. 284–304.

Huglo, M., "Gerbert, théoricien de la musique, vu de l'an 2000." *Cahiers de civilisation médiévale* 43/170 (2000), pp. 143–160.

Juste, D., *Les Alchandreana primitifs. Étude sur les plus anciens traités astrologiques latins d'origine arabe (Xe siècle)*. Leiden: Brill, 2007.

Kárpáti, A., "Translation or Compilation? Contributions to the Analysis of Sources of Boethius' *De institutione musica*." *Studia Musicologica Academiae Scientiarum Hungariae* 29 (1987), pp. 5–33.

Kibre, P., "The Boethian De Institutione Arithmetica and the Quadrivium in the Thirteenth Century University Milieu at Paris." In: Masi, M. (ed.), *Boethius and the Liberal Arts*. Berne – Frankfurt a. M.: Peter Lang, 1981, pp. 67–80.

Klinkenberg, H. M., "Divisio philosophiae." In: Craemer-Ruegenberg, I. – Speer, A. (eds.), *Scientia und ars im Hoch- und Spätmittelalter* I. Berlin: De Gruyter, 1994, pp. 3–19.

Kurth, G., *Notger de Liège et la civilisation au 10ᵉ siècle*. Vol. 1. Paris – Bruxelles – Liège: A. Picard, 1905.

La Croix, R. R. (ed.), *Augustine on Music: An Interdisciplinary Collection of Essays*. Lewiston: Edwin Mellen Press, 1988.

La Rocca, C., "A man for all seasons: Pacificus of Verona and the creation of a local Carolingian past." In: Hen, Y. – Innes, M. (eds.), *The Uses of the Past in the Early Middle Ages*. Cambridge: Cambridge University Press, 2004, pp. 250–277.

Lake, J., *Richer of Saint-Rémi. The Methods and Mentality of a Tenth-Century Historian*. Washington: Catholic University of America Press, 2013.

—, "Gerbert of Aurillac and the Study of Rhetoric in Tenth-Century Rheims." *The Journal of Medieval Latin* 23 (2013), pp. 49–85.

Landes, R., "The Fear of an Apocalyptic Year 1000: Augustinian Historiography, Medieval and Modern." *Speculum* 75/1 (2000), pp. 97–145.

—, "Rudolfus Glaber and the Dawn of the New Millenium: Eschatology, Historiography, and the Year 1000." *Revue Mabillon. Revue Internationale d'Histoire et de Littérature Religieuses* 7 (1996), pp. 57–77.

Lattin, Pratt H., "The origin of our present system of notation according to the theories of Nicholas Bubnov." *Isis* 19/1 (1933), pp. 181–194.

Le Goff, J., *Les Intellectuels au Moyen Âge*. Paris: Éditions du Seuil, 1957.

Leach, E. E., *Sung Birds. Music, Nature, and Poetry in the Later Middle Ages*. Ithaca – London: Cornell University Press, 2007.

Lichfield, M., "Aristoxenus and Empiricism: A Reevaluation Based on His Theories." *Journal of Music Theory* 32/1 (1988), pp. 51–73.

Lindgren, U., *Gerbert von Aurillac und das Quadrivium. Untersuchungen zur Bildung im Zeitalter der Ottonen*. Wiesbaden: Steiner Verlag, 1976.

Lutz, C. E., *Schoolmasters of the Tenth Century*. Hamden, Conn.: Archon Books, 1977.

MacKinney, L. C., *Bishop Fulbert and Education at the School of Chartres*. Notre Dame: University of Notre Dame, 1957.

Marenbon, J., *Boethius*. Oxford: Oxford University Press, 2003.

—, *Early Medieval Philosophy*. London – New York: Routledge, 1988.

Masi, M., "Arithmetic." In: Wagner, D. L. (ed.), *The Seven Liberal Arts in the Middle Ages*. Bloomington: Indiana University Press, 1983, pp. 147–167.

—, "Boethius' De institutione arithmetica in the Context of Medieval Mathematics." In: Obertello, L. (ed.), *Atti del Congresso internazionale di studi Boeziani*. Roma: Editrice Herder, 1981, pp. 263–272.

—, "The Influence of Boethius *De Arithmetica* on Late Medieval Mathematics." In: Masi, M. (ed.), *Boethius and the Liberal Arts: A Collection of Essays*. Berne – Frankfurt a. M., 1981, pp. 81–95.

—, "The Liberal Arts and Gerardus Ruffus' Commentary on the Boethian *De Arithmetica*." *Sixteenth Century Journal* 10/2 (1979), pp. 23–41.

Materni, M., "La *Geometria Gerberti*: un manuale scolastico del X secolo." *Euphrosyne* 37 (2009), pp. 363–374.

—, "Attività scientifiche di Gerberto d'Aurilliac." *Archivum Bobiense* 29 (2007), pp. 225–317.

Mathiesen, T. J., *Apollo's Lyre. Greek Music and Music Theory in Antiquity and the Middle Ages*. Lincoln – London: University of Nebraska Press, 1999.

Matthews, M. R., *Time for Science Education. How Teaching the History and Philosophy of Pendulum Motion Can Contribute to Science Literacy*. Berlin – New York: Springer, 2000.

Mazur, J., *Enlightening Symbols: A Short History of Mathematical Notation and Its Hidden Powers*. Princeton – Oxford: Princeton University Press, 2014.

McCluskey, S. C., "Astronomy in the Service of Christiany." In: Ruggles, C. L. N. (ed.), *Handbook of Archaeoastronomy and Ethnoastronomy*. New York: Springer, 2015, pp. 165–179.

—, *Astronomies and Cultures in Early Medieval Europe*. Cambridge: Cambridge University Press, 1998.

McCready, W. D., "Isidor, the Antipodeans, and the Shape of the Earth." *Isis* 87/1 (1996), pp. 108–127.

McKitterick, R., "Ottonian intellectual culture in the tenth century and the role of Theophano." In: Davids, A. (ed.), *The Empress Theophano: Byzantium and the West at the turn of the first millenium*. Cambridge: Cambridge University Press, 1995, pp 169–193.

Melve, L., "'The revolt of the medievalists'. Directions in recent research on the twelfth-century renaissance." *Journal of Medieval History* 32 (2006), pp. 231–252.

Meyer, C., "*Gerbertus musicus*: Gerbert et les fondements du système acoustique." In: Charbonnel, N. – Iung, J. E. (eds.), *Gerbert l'europeen. Actes du colloque d'Aurillac 4–7 juin 1996.* Aurillac: Société des lettres, sciences et arts La Haute Auvergne, 1997, pp. 183–192.

Meyer, H. – Suntrup, R. (eds.), *Lexikon der mittelalterlichen Zahlenbedeutungen.* München: W. Fink, 1987.

Michel, H., "Les tubes optiques avant le telescope." *Ciel et terre: Bulletin de la societe beige d'astronomie, de meteorologie et de physique du globe* 70 (1954), pp. 175–184.

Miller, G. A., "Gerbert's Letter to Adelbold." *School Science and Mathematicas* 21 (1921), pp. 649–653.

—, "The Formula $1/2a\,(a + 1)$ for the Area of an Equlateral Triangle." *American Mathematical Monthly* 28 (1921), pp. 256–258.

Moretti, G., "Harmonia allegorica: Il melos multiforme che fanda l'armonia del mondo nel De nuptiis Philologiae et Mercurii di Marziano Capella." *PAN. Rivista di Filologia Latina* 2 n.s. (2013), pp. 131–158.

Mostert, M., *The political theology of Abbo of Fleury: a study of the ideas about society and law of the tenth-century monastic reform movement.* Hilversum: Verloren, 1987.

Moyer, A. E., "The *Quadrivium* and the Decline of Boethian Influence." In: Kaylor, N. H. Jr. – Phillips, P. E. (eds.), *A Companion to Boethius in the Middle Ages.* Leiden – Boston: Brill, 2012, pp. 479–517.

—, *The Philosophers' Game. Rithmomachia in Medieval and Renaissance Europe.* Ann Arbor: University of Michigan Press, 2001.

Mülke, M., ""Isidorische Renaissance" – oder: Über die Anbahnung einer Wiedergeburt." *Antiquité Tardive* 23 (2015), pp. 95–107.

Naumann, H., *Karolingische und Ottonische Renaissance.* Frankfurt a. M.: Englert und Schlosse, 1926.

Nelson, J. L., "The Dark Ages." *History Workshop Journal* 63/1 (2007), pp. 191–201.

—, "On the Limits of the Carolingian Renaissance." *Studies in Church History* 14 (1977): *Renaissance and Renewal in Christian History*, pp. 51–69.

Novikoff, A., "Anselm, Dialogue, and the Rise of Scholastic Disputation." *Speculum* 86/2 (2011), pp. 387–418.

—, "The Renaissance of the Twelfth Century Before Haskins." *The Haskins Society Journal. Studies in Medieval History* 16 (2005), pp. 104–116.

Nuvolone, F. G., "Numeri, Croce e Vita: Gerberto e la Parola. A proposito della rilegatura di Echternach: un programma Gerbertiano?." *GERBERTVS – International Academic Publication on History of Medieval Science* 1 (2010), pp. 110–169.

—, "Quelques éléments d'introduction au *Carmen Figuratum* de Gerbert d'Aurillac pubblicato." In: Sigismondi, C. (ed.), *Culmina Romulea: fede e scienza in Gerberto, papa filosofo*. Roma: Ateneo Pontificio Regina Apostolorum, 2008, pp. 47–83.

—, "Gerberto e la musica." *Archivum Bobiense* 5 (2005), pp. 145–164.

—, "Il *Carmen* figurato attribuito a Gerberto nel Ms Paris, BNF, lat. 776, fol. 1v: una composizione redatta nell'Abbazia di San Colombano di Bobbio?." *Archivum Bobiense* 24 (2002), pp. 123–260.

—, "'Gerbertus musicus' e le attività culturali bobbiesi dell'annata 2002." *Archivum Bobiense* 24 (2002), pp. 7–48.

O'Connor, R., "Irish narrative literature and the Classical tradition." In: O'Connor, R. (ed.), *Classical Literature and Learning in Medieval Irish Narrative*. Cambridge: D. S. Brewer, 2014, pp. 1–24.

Oestmann, G., "On the History of the Nocturnal." *Bulletin of the Scientific Instrument Society* 69 (2001), pp. 5–9.

Oldoni, M., *Gloria magistri. Orizzonti letterarî e orali della cultura mediolatina*. Spoletto: Fondazione Centro italiano di studi sull'alto Medioevo, 2018.

—, "'A fantasia dicitur fantasma' (Gerberto e la sua storia, II) II." *Studi Medievali* 24/1 (1983), pp. 167–245

—, "'A fantasia dicitur fantasma' (Gerberto e la sua storia, II)." *Studi Medievali* 21/2 (1980), pp. 493–622.

—, "Gerberto e la sua Storia." *Studi Medievali* 18/2 (1977), pp. 629–704.

Otisk, M., "Gerbert of Aurillac (Pope Sylvester II) as a Clockmaker." *Teorie vědy / Theory of Science* 42/1 (2020), pp. 25–49.

—, "Pojednání Gerberta z Remeše *O sféře*." *Aithér* 11/21 (2019), pp. 26–45.

—, "Gerbert a aritmetika: mezi filosofií čísla a počtářským uměním." *Filosofický časopis* 67/4 (2019), pp. 503–520.

—, "Why Gerbert of Aurillac added to the letter to brother Adam the clime table where the longest day of the year reaches 18 hours?." *GERBERTVS - International Academic Publication on History of Medieval Science* 11 (2018), pp. 1–12 [73–84].

—, "Letter on Timekeeping of Gerbert of Aurillac to Brother Adam." *Constantine's Letters / Konštantínove listy* 11/1 (2018), pp. 67–78.

—, "Dva astronomické dopisy papeže Silvestra II." *Studia theologica* 20/1 (2018), pp. 143–164.

—, "Philosophical Way to God's Wisdom: Arithmetic and the Definitions of a Number in Early Medieval Texts." *Archiwum Historii Filozofii i Myśli Społecznej / Archive of the History of Philsophy and Social Thought* 62 (2017), pp. 55–69.

—, "Ludi philosophorum: filozofowie i zabawy w średniowieczu." In: Teterycz-Puzio, A. –Bejda, W. (eds.), *Homo ludens. Zabawy i rozrywki na co dzień i od święta w dziejach krajów Europy Środkowej i ich sąsiadów*. Słupsk: Wydawnictwo Naukowe Akademii Pomorskiej w Słupsku, 2017, pp. 35–52.

—, "Descriptions and Images of the Early Medieval Latin Abacus." *Średniowiecze Polskie i Powszechne* 7/11 (2015), pp. 13–35.

—, "Rithmomachie – souboj čísel čili aritmetická hra filosofů." *Aithér* 7/14 (2015), pp. 118–139.

—, "Orientacja w czasie i pomiar czasu we wczesnym średniowieczu." *Filozofia. Prace Naukowe Akademii im. Jana Długosza w Częstochowie* 12 (2015), pp. 89–108.

—, "The Philosophical and Mathematical Context of Two Gerbert's Musical Letters to Constantine." *GERBERTVS – International Academic Publication on History of Medieval Science* 8 (2015), pp. 19–38.

—, "The Interpretations and Applications of Boethius's Introduction to the Arithmetic II, 1 at the End of the 10th Century." *GERBERTVS – International Academic Publication on History of Medieval Science* 5 (2014), pp. 33–56.

—, "*Regulae multiplicationis* v abacistických textech Gerberta z Remeše, Abbona z Fleury, Herigera z Lobbes a Bernelia z Paříže." *Pro-Fil* 12/1 (2011), pp. 3–41.

—, *Papežovo ďábelské vědění. Věda a filosofie v době Gerberta z Aurillacu*. Ostrava: Montanex – Ostravská univerzita, 2010.

—, "Gerbert z Aurillacu a abacistické počtářské umění." *Dějiny věd a techniky* 43/2 (2010), pp. 77–102.

—, "Gerbertův úvod do geocentrické astronomie." *Teorie vědy / Theory of Science* 32/3 (2010), pp. 507–533.

—, "Horologium Gerberta z Remeše a časová klimata v raném středověku." *Pro-Fil* 10/2 (2009), pp. 23–44.

—, "Ravennský spor o rozdělení filosofie." *Sborník prací Filozofické fakulty brněnské univerzity* B 53 (2006), pp. 5–19.

Otisk, M. – Psík, R., *Raně středověký latinský abakus. Gerbert z Remeše: Regulae de numerorum abaci rationibus / Pravidla počítání na abaku. Bernelin mladší z Paříže: Liber abaci / Kniha o abaku*. Praha – Ostrava: Scriptorium – Ostravská univerzita, 2020.

—, *Matematické listy Gerberta z Remeše*. Transl. M. Otisk – R. Psík. Praha: Matfyzpress, 2014.

Patzelt, E., *Die Karolingische Renaissance. Beiträge zur Geschichte der Kultur des Frühen Mittelalters*. Wien: Österreichischer Schulbücherverlag, 1924.

Peden, A. M., "Introduction." In: Abbo of Fleury and Ramsey, *Commentary on the Calculus of Vctorius of Aquitaine*. Ed. A. M. Peden. Oxford: Oxford University Press – The British Academy, 2003, pp. xi–liii.

Petrovićová, K., *Martianus Capella. Nauky „na cestě" mezi antikou a středověkem*. Brno: Host, 2010.

Petrucci, F. M., "Theon of Smyrna: Re-thinking Platonic Mathematics in Middle Platonism." In: Tarrant, H. et al. (eds.), *Brill's Companion to the Reception of Plato in Antiquity*. Leiden – Boston: Brill, 2018, pp. 143–155.

Pizzani, U., "Il *Quadrivium* Boeziano e i suoi problem." In: Obertello, L. (ed.), *Atti del Congresso internazionale di studi Boeziani*. Roma: Editrice Herder, 1981, pp. 211–226.

—, "The Fortune of the *De Institutione Muscia* from Boethius to Gerbert D'Aurillac: A Tentative Contribution." In: Masi, M. (ed.), *Boethius and the Liberal Arts: A Collection of Essays*. Berne – Frankfurt a. M., 1981, pp. 97–138.

—, "Studi sulle fonti del "De Institutione Musica" di Boezio." *Sacris erudiri* 16 (1965), pp. 5–164.

Polara, G., "Secolo VII." In: Leonardi, C. (ed.), *Letteratura latina medievale (secoli VI-XV). Un manuale*. Firenze: SISMEL – Edizioni del Galluzzo, 2002, pp. 17–40.

Poulle, E, "Gerbert horloger." In: Guyotjeannin, O. – Poulle, E. (eds.), *Autour de Gerbert d'Aurillac, le pape de l'an mil*. Paris: École des chartes, 1996, pp. 365–367.

Pullan, J. M., *The History of the Abacus*. New York: F. A. Praeger Publishers, 1968.

Radke, G., *Die Theorie der Zahl im Platonismus: Ein systematisches Lehrbuch*. Tübingen: A. Francke, 2003.

Reilly, D. J., "The Bible as Bellwether: Manuscript Bible in the Context of Spiritual, Liturgical and Educational Reform, 1000–1200." In: Poleg, E. – Light, L. (eds.), *Form and Function in the Late Medieval Bible*. Leiden – Boston: Brill, 2013, pp. 9–29.

Ribémont, B., "Isidore de Séville et les mathématiques." In: Baillaud, B. – De Gramont, J. – Hüe, D. (eds.), *Discours et savoirs: encyclopédies médiévales*. Rennes: Presses universitaires de Rennes, 1998, pp. 23–43.

Riché, P., *Abbon de Fleury: un moine savant et combatif (vers 950–1004)*. Turnout: Brepols, 2004.

—, *Gerbert d'Aurillac: Le pape de l'an mil*. Paris: Fayard, 1987.

Riché, P. – Verger, J., *Des nains sur des épaules de géants. Maîtres et élèves au Moyen Âge*. Paris: Tallandier, 2006.

Rimple, M. T., "The Enduring Legacy of Boethian Harmony." In: Kaylor, N. H. Jr. – Phillips, P. E. (eds.), *A Companion to Boethius in the Middle Ages*. Leiden – Boston: Brill, 2012, pp. 447–478.

Rollo, D., *Kiss My Relics: Hermaphroditic Fictions of the Middle Ages*. Chicago – London: University of Chicago Press, 2011.

Roskam, G., *On the Path to Virtue. The Stoic Doctrine of Moral Progress and its Reception in (Middle-)Platonism*. Leuven: Leuven University Press, 2005.

Rossi, P., "Sinossi delle principali differenti proposte di datazione." In: Gerbert, *Lettere (983–997)*. Transl. P. Rossi. Pisa: Plus – Pisa University Press, 2009, pp. 201–205.

Rouche, M. (ed.), *Fulbert de Chartres. Précurseur de l'Europe médiévale?*. Paris: Presses de l'Univeristé Paris-Sorbonne, 2008.

Rowett, C., "Philosophy's Numerical Turn: Why the Pythagoreans' Interest in Numbers is Truly Awesome." In: Sider, D. – Obbink, D. (eds.), *Doctrine and Doxography. Studies on Heraclitus and Pythagoras*. Berlin – Boston: De Gruyter, 2013, pp. 3–31.

Rückert, O., *Ottonische Renaissance. Ausgewählte Stücke aus Widukind von Corvey, Ruotger, Liudprand von Cremona, Hrotsvit von Gandersheim, Ekkehard IV. von St. Gallen*. Leipzig: Teubner, 1926.

Sachs, K. J., "Gerbertus cognomento musicus. Zur musikgeschichtlichen Stellung des Gerbert von Reims (nachmaligen Papstes Silvester II)." *Archiv für Musikwissenschaft* 29 (1972), pp. 257–274.

Saenger, P., *Space between Words. The Origins of Silent Reading*. Stanford: Stanford University Press, 1997.

Saliba, G., *Islamic Science and the Making of the European Renaissance*. Cambridge: MIT Press, 2007.

Samsó, J., "Cultura científica àrab i cultura científica latina a la Catalunya altmedieval: El monestir de Ripoll i el naixement de la ciència catalna." In: Udina i Martorell, F. (ed.), *Symposium internacional sobre els oríges de Catalunya (segles VIII–XI)*. Vol. 1. Barcelona: RABL, 1991, pp. 253–269.

Savage-Smith, E., *Islamicate Celestial Globes: Their History, Construction, and Use*. Washington: Smithsonian Institution Press, 1985.

Schärling, A., *Un portrait de Gerbert d'Aurillac: Inventeur de l'abaque, utilisateur précoce des chiffres arabes et pape de l'an mil*. Lausanne: Presses polytechniques et universitaires romandes, 2012.

Schmid, H., "Zur sogenannten 'Musica Adelboldi Traiectensis'." *Acta Musicologica* 28/2 (1956), pp. 69–73.

Schrade, L., "Music in the Philosophy of Boethius." *The Musical Quarterly* 33/2 (1947), pp. 188–200.

Schulmeyer-Ahl, K., *Der Anfang vom Ende der Ottonen. Konstitutionsbedingungen historiographischer Nachrichten in der Chronik Thietmars von Merseburg*. Berlin – New York: De Gruyter, 2009.

Shelby, L. R., "The geometrical knowledge of mediaeval master masons." In: Courtenay, L. T. (ed.), *The Engineering of Medieval Cathedrals*. London – New York: Routledge, 1997, pp. 27–61.

—, "Geometry." In: Wagner, D. L. (ed.), *The Seven Liberal Arts in the Middle Ages*. Bloomington: Indiana University Press, 1983, pp. 196–217.

Sigismondi, C., "Gerberto, gli Arabi e Gerusalemme." *GERBERTVS – International Academic Publication on History of Medieval Science* 1 (2010), pp. 270–294.

—, "Gerberto e la misura delle canne d'organo." *Archivum Bobiense* 29 (2007), pp. 355–396.

Sigismondi, F., *Gerberto d'Aurillac, il trattato De Rationali et Ratione Uti e la Logica del X secolo*. Roma: Ateneo Pontificio Regina Apostolorum, 2007.

Silva, J. N., "O Ábaco de Gerbert." *GERBERTVS – International Academic Publication on History of Medieval Science* 4 (2013), pp. 102–119.

—, "Teaching and playing 1000 years ago, Rithmomachia." In: Sigismondi, C. (ed.), *Orbe novus. Astronomia e Studi Gerbertiani*. Roma: Universitalia, 2010, pp. 135–148.

—, "Mathematical games in Europe around the year 1000." *GERBERTVS – International Academic Publication on History of Medieval Science* 1 (2010), pp. 213–225.

Šíma, A., *Svět vymezený a neomezený. Principy přírody ve filosofii Filoláa z Krotónu a u raných pythagorejců*. Červený Kostelec: Pavel Mervart, 2012.

Smith, D. E., *History of Mathematics. Vol. II: Special Topics of Elementary Mathematics*. New York: Dover Publications, 1958.

Špelda, D., *Astronomie ve středověku*. Ostrava: Montanex, 2008.

Stahl, W. H., *The Quadrivium of Martianus Capella: Latin Tradition in Mathematical Sciences 50 B.C.–A.D. 1250. Martianus Capella and the Seven Liberal Arts*. Vol. 1. New York – London: Columbia University Press, 1971.

—, *Roman Science. Origins, Development, and Influence to the Later Middle Ages*. Madison: University of Wisconsin Press, 1962.

Stautz, B., "Die früheste bekannte Formgebung der Astrolabien." In: von Gotstedter, A. (ed.), *Ad radices. Festband zum fünfzigjährigen Bestehen des Instituts für Geschichte der Naturwissenschaften der Johann Wolfgang Goethe-Universität, Frankfurt am Main*. Stuttgart: Steiner Verlag, 1994, pp. 315–328.

Stevens, W. M. – Beaujouan. G. – Turner, A. J. (eds.), *The Oldest Latin Astrolabe. Physis: Rivista internazionale di storia della scienza* 32 (1995).

Studtmann, P., "Aristotle's Category of Quantity: A Unified Interpretation." *Apeiron* 37/1 (2004), pp. 69–91.

Sugden, K. F., "A History of the Abacus." *Accounting Historians Journal* 8/2 (1981), pp. 1–22.

Swanson, R. N., *The Twelfth-Century Renaissance.* Manchester: Manchester University Press, 1999.

Teeuwen, M. "The Pursuit of Secular Learning: The Oldest Commentary Tradition on Martianus Capella." *Journal of Medieval Latin* 18 (2008), pp. 36–51.

—, *Harmony and the Music of the Spheres. The Ars Musica in Ninth-Century Commentaries on Martianus Capella.* Leiden: Brill, 2002.

Thompson, R. M. – Winterbottom, M., *William of Malmesbury, The History of the English Kings: General Introduction and Commentary.* Oxford: Clarendon Press, 1999.

Thomson, R. B., "Further Astronomical Material of Abbo of Fleury." *Mediaeval Studies* 50 (1988), pp. 671–673.

—, "Two Astronomical Tractates of Abbo of Fleury." In: North, J. D. – Roche, J. J. (eds.), *The Light of Nature. Essays in the History and Philosophy of Science presented to A. C. Crombie.* Dordrecht – Boston – Lancaster: Martinus Nijhoff Publishers, 1985, pp. 113–133.

Thorndike, L., "Invention of the Mechanical Clock about 1271 A.D." *Speculum* 16/2 (1941), pp. 242–243.

Tosi, M. (ed.), *Gerberto – scienza, storia e mito. Atti del Gerberti Symposium.* Bobbio: A.S.B., 1985.

Trompf, G. W., "The Concept of the Carolingian Renaissance." *Journal of the History of Ideas* 34/1 (1973), pp. 3–26.

Truitt, E. R., *Medieval Robots. Mechanism, Magic, Nature, and Art.* Philadelphia: University of Pennsylvania Press, 2015.

—, "Celestial Divination and Arabic Science in Twelfth-Century England: The History of Gerbert of Aurillac's Talking Head." *Journal of the History of Ideas* 73/2 (2012), pp. 201–222.

Udina i Martorell, F., "Gerberto y la cultura hispanica: los Manuscriots de Ripoll." In: Tosi, M. (ed.), *Gerberto – scienza, storia e mito. Atti del Gerberti Symposium.* Bobbio: A.S.B., 1985, pp. 35–50.

Usher, A. P., *A History of Mechanical Inventions. Revised Edition.* New York: Dover Publications, 1982.

Van de Vyver, A., "Les plus anciennes Traductions latines médiévales (X^e-XIe siècles) de Traités d'Astronomie et d'Astrologie". *Osiris* 1 (1936), pp. 658-691.

—, "Les œuvres inédites d'Abbon de Fleury." *Revue Bénédictine* 47 (1935), pp. 125-169.

Vasina, A., "Gerberto arcivescovo di Ravenna." In: Tosi, M. (ed.), *Gerberto – scienza, storia e mito. Atti del Gerberti Symposium*. Bobbio: A.S.B., 1985, pp. 255-272.

Vaughn, S. N. – Rubenstein, J. (eds.), *Teaching and Learning in Northern Europe, 1000-1200*. Turnhout: Brepols, 2006.

Vogel, C., "Die „boethianische Frage" – Über die Eigenständigkeit von Boethius' logischem Lehrwerk." *Working Paper des SFB 980 Episteme in Bewegung* 17 (2019), pp. 1-28.

—, *Boethius' Übersetzungsprojekt. Philosophische Grundlagen und didaktische Methoden eines spätantiken Wissenstransfers*. Wiesbaden: O. Harrasowitz, 2016.

Walden, D. K., "Charting Boethius: Music and the Diagrammatic Tree in the Cambridge University Library's *De Institutione Arithmetica*, MS II.3.12." *Early Music History* 34 (2015), pp. 207-228.

Wallis, F., "Isidore of Seville and Science." In: Fear, A. – Wood, J. (eds.), *A Companion to Isidore of Seville*. Leiden: Brill, 2019, pp. 182-221.

Wangerin, L. E., *Kingship and Justice in the Ottonian Empire*. Ann Arbor: University of Michigan Press, 2019.

Warren, F. M., "Constantine of Fleury, 985-1014". *Transactions of the Connecticut Academy of Arts and Science* 15 (1909), pp. 283-292.

Wedell, M., "Numbers". In: Classen, A. (ed.), *Handbook of Medieval Culture. Fundamental Aspects and Conditions of the European Middle Ages*. Vol. II. Berlin – Boston: De Gryuter, 2015, pp. 1205-1260.

Weisheipl, J. A., "The Nature, Scope, and Classification of the Sciences." In: Lindberg, D. C. (ed.), *Science in the Middle Ages*. Chicago: University of Chicago Press, 1977, pp. 461-482.

—, "The Concept of Scientific Knowledge in Greek Philosophy." In: Gagne, A. – De Koninck, T. (eds.), *Melanges a la Memoire de Charles De Koninck*. Quebec: Les Presses de l'Université Laval, 1968, pp. 487-507.

Wessely, O., "Adelbold von Utrecht und seine 'Musica'." *Anzeiger der Kaiserlichen Akademie der Wissenschaften. Philosophisch-historische Klasse* 24 (1949), pp. 575-583

Wiesenbach, J., "Der Mönch mit dem Sehohr. Die Bedeutung der Miniatur Codex Sangallensis 18, p. 45." *Schweizerische Zeitschrift für Geschichte* 44/4 (1994), pp. 367-388.

—, "Pacificus von Verona als Erfinder einer Sternenuhr." In: Butzer, P. L. – Lohrmann, D. (eds.), *Science in Western and Eastern civilization in Carolingian times*. Basel: Birkhäuser Verlag, 1993, pp. 229–250.

Williams, J. R., "The Cathedral School of Rheims in the Eleventh Century." *Speculum* 29/4 (1954), pp. 661–677.

Williams, P., *The Organ in Western Culture, 750–1250*. Cambridge: Cambridge University Press, 1993.

Wood, I., "Culture." In: McKitterick, R. (ed.), *The Early Middle Ages. Europe 400–1000*. Oxford: Oxford University Press, 2001, pp. 167–198.

Zimmermann, H., "Gerbert als kaiserlicher Rat." In: Tosi, M. (ed.), *Gerberto – scienza, storia e mito. Atti del Gerberti Symposium*. Bobbio: A.S.B., 1985, pp. 235–253.

Zuccato, M., "Gerbert of Aurillac and a Tenth-Century Jewish Channel for the Transmission of Arabic Science to the West." *Speculum* 80 (2005), pp. 742–763.

INDICES

List of Tables

Table 1	– Ten types of inequality	39
Table 2	– The emergence of inequalities from equality	41
Table 3	– Boethius's example of converting multiples to equality according to [R1–3] rules	43
Table 4	– Notker's approach for conversion of a 5 : 4 ratio to equality	46
Table 5	– The conversion of superparticular multiples to equality according to Notker	48
Table 6	– The conversion of superpartient multiples to equality according to Notker	48
Table 7	– The so-called *Saltus Gerberti*, i.e. the transition between superparticular ratios and multiples	49
Table 8	– Forming the sequence of triangular numbers	57
Table 9	– Forming the sequence of plane numbers from triangular numbers	58
Table 10	– Gerbert's horological tables (Letter to Adam)	105
Table 11	– Daylight according to Gerbert's two horological tables	112
Table 12	– Symbols and names of *ghubar*-numerals	148
Table 13	– Markings of the abacus columns	152
Table 14	– Fractions according to drawings of early medieval abacuses	155
Table 15	– Multiplication table according to Bernelin's *Liber abaci*	166
Table 16	– Different orderings of Gerbert's rules of division	172

List of Figures

Fig. 1 – Rithmomachy, the starting position of chips 54
Fig. 2 – Triangular basis of plane numbers 56
Fig. 3 – Triangular numbers 57
Fig. 4 – Point visualisation of the arithmetical equation for the area of an equilateral triangle, showing the necessity of adding the length of a side to its square root 63
Fig. 5 – Length and superficial measurement *versus* linear and plane number 66
Fig. 6 – Triangular number 28 according to arithmetic, i.e., the arithmetical visualisation of an equilateral triangle with a side 7 feet long 66
Fig. 7 – Equilateral triangle with a side 7 feet long according to geometry 67
Fig. 8 – Calculating the area of an equilateral triangle according to geometry via a sum of square feet delineated by sides 68
Fig. 9 – Subtraction of two consequent superparticular ratios according to Gerbert's schemes 73
Fig. 10 – Duplex of the superparticular ratio a 3 : 2 gives the result of the multiplex superparticular ratio 9 : 4; based on Gerbert's example 74
Fig. 11 – Duplex of the difference between a perfect fifth and a perfect fourth is less than a perfect fourth; based on Gerbert's example 77
Fig. 12 – Celestial sphere 87
Fig. 13 – Horizon and astronomical meridian 89
Fig. 14 – Orbits of stars observable over the horizon 90
Fig. 15 – Five parallel circles of the world sphere 92
Fig. 16 – Gerbert's placement of the five parallels of the world sphere according to the letter addressed to Constantine *De sphaera* 94
Fig. 17 – Gerbert's observational hemisphere, including an iron semicircle for fixing sevenobservational tubes 96
Fig. 18 – Ecliptic, zodiac, and zodiacal signs 98
Fig. 19 – The colure of the equinox (*colur aequinoctialis*) and the colure of the solstice (*colur solstitialis*) 100
Fig. 20 – Daylight in the climate of Hellespont according to the theory of equal changes and according to the theory of unequal changes 108
Fig. 21 – Division of the Earth 109

Fig. 22 – A celestial sphere with the Earth in its centre and its image on the front side of an astrolabe 116
Fig. 23 – Determining the current time using the astrolabe 120
Fig. 24 – Possible appearance of clocks ascribed to Pacificus of Verona 125
Fig. 25 – Bernelin's abacus according to *Liber abaci* I 136
Fig. 26 – Abacus from Echternach (before 1000); drawing according to Luxemburg, Bibliothèque nationale de Luxembourg, MS 770 and Trier, Stadtbibliothek, MS 1093/1694, fol. 197r 137
Fig. 27 – The Bern abacus (end of 10th century); drawing according to Bern, Burgerbibliothek, MS 250, fol. 1r 139
Fig. 28 – The Paris abacus (beginning of 11th century); drawing according to Paris, Bibliothèque nationale de France, Lat. 8663, fol. 49v 140
Fig. 29 – Abacus from the so-called Pseudo-Boethius's *Geometria II* (first half of 11th century); drawing according to Erlangen, Universitätsbibliothek 379, fol. 35r 141
Fig. 30 – The Vatican abacus (11th century); drawing according to Vatican, Lat. 644, fols. 77v–78r 143
Fig. 31 – The Rouen abacus (11th century); drawing according to Rouen, Bibliothèque municipale, MS 489, fols. 68v–69r 144
Fig. 32 – The second Paris abacus (11th century); drawing according to Paris, Bibliothèque nationale de France, Lat. 7231, fol. 85v 146
Fig. 33 – The Oxford abacus (around the year 1110); drawing according to Oxford, St. John's College, MS 17, fols. 48v–49r 147
Fig. 34 – Gerbert's figurative panegyric poem for Otto II 149
Fig. 35 – Gerbert's figurative panegyric poem for Otto II as a grid poem with acrostic 150
Fig. 36 – Expressing numerical values by fingers, redrawn according to London, British Library, Royal MS 13 A XI, fol. 33v 161
Fig. 37 – Multiplication according to Bernelin's *Liber abaci* I 169
Fig. 38 – Division according to Gerbert's 1st rule 173
Fig. 39 – Division according to Gerbert's 2nd rule 175
Fig. 40 – Division according to Gerbert's 4th rule 177
Fig. 41 – Division according to Gerbert's 5th rule 179
Fig. 42 – Division according to Gerbert's 9th rule 182

Index locorum

A
Abbo of Fleury
- *Abacus*
  203–204: 133, 164
- *De quinque circulus mundi* [= *De circ.*]
  672–673: 94
- *De ratione spere* [= *De spere*]
  120: 86, 88, 97, 99, 101
  121: 97
- *Explanatio in Calculo Victorii* [= *In Calc.*]
  I, 1–3: 17
  II, 1: 17, 36, 42
  II, 2: 17
  II, 3: 17
  II, 4–6: 17
  II, 7: 17
  II, 8–9: 17
  II, 10–11: 17, 42
  II, 12–16: 17
  III, 1: 42
  III, 2: 42, 51
  III, 3: 42
  III, 5: 42
  III, 22: 36, 37, 50
  III, 23: 37, 50
  III, 24: 36, 50
  III, 25–26: 50
  III, 27: 50
  III, 37: 122
  III, 64: 133, 164
  III, 65–66: 133
  III, 67: 133, 161
  III, 68: 162
  III, 73: 38
  III, 74: 38
  III, 75–76: 38
  III, 77–78: 38
  III, 79: 38
  III, 83: 51
- *Excerpta in calculum Victorii commentraio* [= *Excerpta*]
  197–202: 133
  203: 158
- *Quaestiones grammaticales* [= *Quaest. gram.*]
  50: 36
Adelard of Bath
- *Regule abaci* [= *Reg. abaci*]
  91: 131, 132, 180
Ademar of Chabannes
- *Chronicon* [= *Chron.*]
  III, 31: 35, 83
Annales regni Francorum [= *Ann. Franc.*]
- 807: 122
Aristotle
- *Atheniensium respublica* [= *Ath. pol.*]
  68, 2–4: 129
  69, 1: 130
- *Categoriae* [= *Cat.*]
  6: 20, 26
- *De anima* [= *De an.*]
  III, 7: 19
- *Metaphysica* [= *Met.*]
  I, 5: 24, 69
  V, 13: 20, 26
  VI, 1: 19
  X, 1: 23
  XI, 7: 18
- *Physica* [= *Phys.*]
  II, 2: 19
- *Topica* [= *Top.*]
  I, 5: 26

Ascelin of Augsburg
- *Compositio astrolabii* [= *Comp. astrol.*]
    3: 107
    Star table: 87
Asilo of Würzburg
- *Regula de rithmachia* [= *Rith.*]
    1: 53
    2: 54
    3–5: 54
    8: 55
Augustine
- *De civitate dei* [= *Civ. Dei*]
    VIII, 10: 186
    XI, 30: 27
    XI, 31: 28
- *De Genesi ad litteram* [= *Gen. ad litt.*]
    II, 1: 23
- *De libero arbitrio* [= *Lib. arb.*]
    II, 8, 20–12, 34: 15
    II, 16, 42: 15, 82
- *De musica* [= *Mus.*]
    I, 2: 78
    III, 2: 23
- *De ordine* [= *Ord.*]
    II, 14, 39: 78
    II, 14, 40: 23, 78
    II, 14, 41: 78
    II, 15, 42: 82
    II, 18, 48: 15

B
Bede the Venerable
- *De natura rerum* [= *De nat. rerum*].
    3: 86
    5: 86
    9: 94
    12: 98
    13: 98
    16–17: 97
- *De temporum ratione* [= *De temp. rat.*]
    1: 160
    5: 106
    33: 111
- *Historia ecclesiastica* [= *Hist. eccles.*]
    I, 1: 111
Bernard of Clairvaux
- *Epistolae* [= *Ep.*]
    385: 185
Bernelin of Paris
- *Liber abaci* [= *Liber abaci*]
    Praef.: 132, 134
    I: 131, 134–136, 152–154, 162, 164–169
    II: 170, 171, 180
    III: 170, 171
Biblia sacra
- *Epistula ad Romanos* [= *Rom.*]
    11, 36: 42
- *Genesis* [= *Gn*]
    1, 3–5: 106
- *Liber Sapientiae* [= *Sap.*]
    11, 20: 23, 42
- *Proverbia* [= *Prov.*]
    9, 1: 42
Boethii quae dicitur geometria altera [= *Geom. II*].
    IV, 2–6: 60
    V, 1–3: 62
    XV, 1: 132
    XX, 1: 132
    XX, 5: 132
Boethius
- *De arithmetica* [= *Arith.*]
    I, 1: 15, 20–23
    I, 2: 26, 37
    I, 3: 24, 26, 28, 29
    I, 4–8: 26
    I, 9–10: 26, 29
    I, 11–16: 26

I, 17: 23, 26
I, 18–19: 26
I, 20: 26, 27
I, 21: 30, 37
I, 22: 30, 38
I, 23: 30, 38
I, 24: 30, 38
I, 25: 30
I, 26: 30, 165
I, 27: 30
I, 28: 30, 38
I, 29: 30, 38
I, 30: 30, 38
I, 31: 30
I, 32: 24, 30, 37, 42, 44
II, 1: 23, 30, 37, 43, 44, 50
II, 2–3: 30
II, 4: 25, 55, 65
II, 5: 25, 65
II, 6: 25, 55, 56
II, 7–9: 25
II, 10–19: 25
II, 20: 25, 65
II, 21–23: 25, 57
II, 24–39: 25
II, 40–47: 30
II, 48: 30, 51
II, 49–50: 30
– *De consolatione philosophiae* [= *De cons. phil.*]
I, 4: 114
– *De institutione musica* [= *Mus.*]
I, 1: 80, 81
I, 9: 81
I, 10: 51
I, 16–19: 51
II, 10: 71
II, 21: 72
– *De sancta trinitate* [= *Trin.*]
2: 19
5: 37

– *In Categorias Aristotelis Commentaria* [= *In Cat.*]
II: 20
– *In Isagogen Porphyrii commentorum editionis primae* [= *1 In Isag.*]
I, 3: 19, 103
Byrhtfert of Ramsey
– *De loquela per gestum digitorum et temporum ratione libellus* [= *De loquela*]
688C–D: 160

C
Calcidius
– *Commentarius in Platonis Timaeum* [= *In Tim.*]
I, 32: 42
II, 326: 58
Cassiodore
– *Institutiones* [= *Inst.*]
I, 30: 122
II, praef.: 20
II, 3: 19
II, 4: 19, 23, 24, 26
II, 5: 37, 38, 78
II, 6: 59
II, 7: 88, 98, 110
– *Variae* [= *Var.*]
I, 45: 19, 122
Censorinus
– *De die natali liber* [= *Die nat.*]
X, 3: 78
Ciacono, Alphons – Oldoini, Augustine
– *Vitae, et res gestae pontificum romanorum et S.R.E. cardinalium ab initio nascentis Ecclesiae usque ad Clementem IX. P.O.M.* [= *Vitae pontif. card.*]
751A–754D: 121

756A: 121
Cicero, Marcus Tullius
- *Tusculanae disputationes* [= *Tusc. disp.*]
  V, 24, 68: 186
Commentarii in Gerberti regulas de numerorum abaci rationibus [= *In Gerb.*]
  I, 4: 131
  II, 1: 159
  II, 2: 160, 162
Consuetudines Floriacenses antiquiores [= *Consuet. Flor.*]
  42: 122

D
De arithmetica Boetii [= *De arith. Boeth.*]
  ad prol.: 16
  ad I, I: 21, 22, 37, 51
  ad I, IIII–XX: 27, 30
  ad I, XXI–XXII: 30
  ad I, XXIII–XXVII: 30
  ad I, XXVIII–XXXI: 30
  ad I, XXXII: 30, 42, 44
  ad II, I: 30
  ad II, II: 30, 51
  ad II, IIII–XXXII: 25
  ad II, XL–LII: 30
De minutiis
  225–244: 133
De mundi coelestis terrestrisque constitutione liber [= *De mundi*]
  883D–884A: 107
Diogenes Laertius
- *Vitae philosophorum* [= *Vitae*]
  VII, 39–40: 186

E
Epaphroditus – Vitruvius Rufus [?]
- *Excerpta*
  4 (II, 1): 64

Epistolae quorundam [= *Ep. quord.*]
  I, 1: 131
Epitaphium Pacifici archidiaconi [= *Epit. Pacif*]
  12–13: 123
  15: 123
Eratosthenes of Cyrene
- *Die geographischen Fragmente des Eratosthenes* [= *Fragm.*]
  3A, 18–40: 110
Eriugena, John Scotus
- *Annotationes in Marcianum* [= *In Marc.*]
  295, 5: 112
  296, 5: 111
- *Periphyseon (De divisione naturae)* [*De div.*]
  III, 33 [717–719]: 86
Euclid
- *Elementa* [= *Elem.*]
  I, prop. 1: 131
  VII, def. 2: 24
  VII, def. 3–4: 30
  VII, def. 6–14: 27
  VII, def. 8: 28
  VII, def. 9: 29
  VII, def. 16–20: 25
  VII, def. 21: 30
  VII, def. 22: 27
  XIII, 13–17: 56
Excerptiuncula [= *Excerpt.*]
  279: 21, 24

G
Geometria incerti auctoris [= *Geom. inc.*]
  IV, 10: 60
Gerbert of Aurillac/of Reims
- *Ad Adelboldum de causa diversitatis arearum trigoni aequlateri geometrice arithmeticeve expensi* [= *Ad Adel.*]

1: 60, 61
2: 65
3: 65–68
– *De sphaera*
  1: 94
  2: 92, 93, 95, 126
  3: 92, 96, 97, 126
– *De utilitatibus astrolabii* [= *De util. astrol.*]
  1, 2: 117
  5, 4: 117
  6, 1: 117
  8, 2: 106
  8, 3: 106, 107
  9–10: 107
  13, 1–2: 112
  21: 118
– *Epistolae* [= *Ep.*]
  7: 85
  8: 85
  17: 157
  24: 83
  25: 83, 157
  44: 82, 185, 186
  92: 51, 70, 82
  112: 83
  130: 85
  134: 91, 115
  148: 91, 115
  153: 103–106, 112
  167: 85
  168: 85
  183: 156
  194: 185
– *Omnigenum pater* [= *Figur.*]
  666–667: 149
– *Fragmentum de norma rationis abaci* [= *Fr. de abaci*]
  23: 186
  24: 159, 186
– *Geometria* [= *Geom.*]
  I, 1–2: 59

I, 2–3: 59
II, 1: 59
II, 5: 65
II, 6: 65, 85
V, 1: 56
V, 1–VII, 1: 61
– *Regulae de numerorum abaci* [= *Regulae*]
  Praef.: 121, 156, 159, 186
  I: 163, 165
  II, 1 [2]: 170, 171, 174, 175
  II, 1 [3]: 171, 176
  II, 2 [4]: 170, 171, 176–178
  II, 2 [5]: 171, 178, 179
  II, 2 [6]: 171, 180
  II, 3 [1]: 173
  II, 4 [7]: 171, 180
  II, 5 [8–9]: 181, 182
  II, 6 [10]: 183
– *Rogatus a pluribus* [= *Rogatus*]
  59–60: 51, 71
  61–81: 51
– *Scholium ad Boethii Arithmeticam Institutionem l. II, c. I* [= *In Boeth. Arith.*]
  1: 36, 44, 49
  2: 49
– *Scholium ad Boethii Musicae Institutionis l. II, c. 10; l. IV, c. 2* [= *1 In Boeth. Mus.*]
  29: 72
– *Scholium ad Boethii Musicae Institutionis l. II, c. 21* [= *2 In Boeth. Mus.*]
  30: 51
  31: 51, 72, 73
Glosses to Boethius' De Arithmetica in Cologne Ms. 186 [= *In Boeth. Arith.*]
  15: 20
  15–40: 27
  41–63: 30

64–81: 25
82–112: 30

H
Heriger of Lobbes
– *Regulae de numerorum abaci rationibus* [= *Regulae*]
  209: 162, 164
– *Regula de abaco computi* [= *Regula*]
  311–324: 163
Hermann of Reichenau
– *De mensura astrolabii* [= *De mens. astrol.*]
  3: 107
  6: 87
– *Regula de rithmimachia* [= *Rith.*]
  1: 53
  2: 54
  3: 54
  4: 55
Hyginus, Gaius Julius
– *De astronomia* [= *De astron.*]
  I, 7: 95

I
Iamblichus of Chalcis
– *In Nicomachi Arithmeticam Introductionem liber* [= *In Nic. Arith.*]
  10: 24, 26
Isidore of Sevilla
– *Etymologiarum sive Originum libri XX* [= *Etym.*]
  I, 17: 52
  II, 24: 186
  III, praef.: 20
  III, 1: 23
  III, 3: 24
  III, 4: 15
  III, 5: 28, 29
  III, 6: 30, 37, 38
  III, 10: 59
  III, 11: 58
  III, 12: 58
  III, 15: 78
  III, 17: 52, 80
  III, 29: 86
  III, 31: 86
  III, 32: 86
  III, 33: 86
  III, 34: 86
  III, 36–37: 86
  III, 42: 110
  III, 43: 88
  III, 44: 94
  III, 50: 97, 106
  III, 51: 99, 106
  III, 52: 106
  III, 64: 98
  XIV, 2: 110

M
Macrobius
– *Commentarii in Somnium Scipionis* [= *In Somn.*]
  I, 6, 46: 42
  I, 15, 16: 88
  I, 15, 17: 88
  I, 20, 20: 86
  I, 21, 12–21: 112
  I, 21, 24: 101
  II, 5, 13–14: 109
Martianus Capella
– *De nuptiis Philologiae et Mercurii* [= *De nupt.*]
  VI, 579: 131
  VI, 580: 59
  VI, 582: 131
  VI, 583–584: 86
  VI, 587: 131
  VI, 588: 59

VI, 595: 111
VI, 724: 131
VII, 725: 131
VII, 728: 29
VII, 729: 29, 131, 160
VII, 743: 24, 29
VII, 761: 38
VII, 762: 38
VIII, 814: 86
VIII, 815: 89
VIII, 827: 95
VIII, 832–833: 99
VIII, 834–835: 97, 106
VIII, 836: 89
VIII, 838: 87
VIII, 849: 101
VIII, 850: 98
VIII, 851: 98
VIII, 855: 101
VIII, 876: 111
VIII, 877: 111
VIII, 878: 104, 108
IX, 888–898: 79
IX, 905: 79
IX, 910: 79
IX, 930: 78
IX, 996: 79

N
Nicomachus of Gerasa
– *Introductio arithmeticae* [= *Arith.*]
  I, 2: 19, 20
  I, 3: 21
  I, 4: 22
  I, 6: 37
  I, 7: 24, 26, 28
  I, 8–9: 26, 29
  I, 10–13: 26
  I, 14–15: 26, 27
  I, 16: 26
  I, 17: 30, 37, 38

I, 18: 30, 38
I, 19: 30, 38
I, 20–21: 30, 38
I, 22: 30, 38
I, 23: 30, 37, 38, 41
II, 1: 30, 37, 43
II, 2: 30, 43
II, 3–5: 30
II, 6: 25, 55
II, 7: 25, 56
II, 8: 25
II, 9–12: 25
II, 13–14: 25, 57
II, 15–20: 25
II, 21–27: 30
Notker of Liège
– *De superparticulari* [= *Superpart.*]
  297: 36, 45
  298: 36, 45
  299: 36, 47

P
Pacificus of Verona
– *Spera caeli* [= *Spera*]
  692: 124
Peter Abelard
– *Carmen ad Astralabium* [= *Ad Astral.*]
  907: 185
Plato
– *Gorgias* [= *Gorg.*]
  451a–c: 15
– *Respublica* [= *Resp.*].
  VII, 6: 23
  VII, 8: 15
– *Theaetetus* [= *Tht.*]
  198a–c: 15
– *Timaeus* [= *Tim.*]
  34b–36d: 42
  53c: 58
  53c–55c: 56

54d–55c: 58
55c–56c: 56
Pliny the Elder
- *Naturalis historiae libri XXXVII* [= *Nat. hist.*]
  II, 64: 101
  VI, 33–34: 111
Ptolemy
- *Syntaxis mathematica* [= *Alm.*]
  I, 1: 22
  II, 12: 110

Q
Quid sit abacus [= *Abacus*]
  625: 131

R
Ralph of Laon
- *Liber de abaco* [= *De abaco*]
  100: 132
Regulae arthimeticae [= *Reg. arith.*]
  608: 131
Regulae domni Odonis super abacum [= *Odon. super abacum*]
  807A: 160
  807B: 132
  807C: 132
  807D: 132
Remigius of Auxerre
- *Commentum in Martianum Capellam* [= *In Marc.*]
  VI, 288: 131
  VII, 363: 131
Richer of Reims
- *Historiarum libri IIII* [= *Hist.*]
  III, 21–23: 103
  III, 43: 35, 83
  III, 44: 70
  III, 46: 82
  III, 47: 82
  III, 48: 82
  III, 49: 51, 75, 83
  III, 50: 86, 88, 90, 91
  III, 51: 92, 93, 97, 126
  III, 52: 99–101
  III, 53: 102
  III, 54: 83, 115, 131, 134
  III, 55: 34
  III, 56: 186
  III, 57: 187
  III, 60: 186
  III, 61: 186
Stobaeus, John
- *Eclogae physicae et ethicae* [= *Ecl.*]
  I, 1, 8: 29

T
Thietmar of Merseburg
- *Chronicon* [= *Chron.*]
  IV, 16: 123
  VI, 100: 93, 122
Tractatus de divisione [= *Tract. de div.*]
  630: 132
Turchill Compotista
- *Reguncule super abacum* [= *Reg. super abacum*]
  135: 130, 132

V
Victorinus of Aquitaine
- *Calculus* [= *Calc.*]
  4–37: 165
  48: 167

W
William of Malmesbury
- *Gesta regum Anglorum* [= *Gesta reg.*]
  II, 167, 1: 35, 84, 113
  II, 167, 2–3: 35, 84, 114
  II, 167, 4–5: 114

II, 168, 1: 114
II, 168, 5: 114
II, 168, 6: 114
II, 169, 2–3: 114
II, 172, 2: 114

Wyon, Arnold
– *Lignum vitae, ornamentum et decus ecclesiae* [= *Lignum vitae*]
V: 121

Index of Personal Names (before 1700)

A
Abbo of Fleury (Abbo Floriacensis) 11–13, 17, 33, 35–38, 42, 48, 50, 51, 53, 86, 88, 94, 95, 97, 99, 101, 122, 132, 133, 158, 161, 162, 164
Adalbero of Reims (Adalbero Remensis) 103, 114
Adam, addressee of Gerbert's letter 34, 81, 103–105, 107, 109, 111–113, 122, 126
Adelard of Bath (Adelardus Bathoniensis) 131–133, 180
Ademar of Chabannes (Ademarus Cabannensis) 35, 83
Adelbold of Utrecht (Adelboldus Traiectensis) 33, 60, 65
Alchandreus (Philosophus) 35, 84
Al-Kindi (Alkindus) 84
Anonymous, author called [Pseudo-]Boethius 60, 133, 140, 141
Aprofiditus *see* Epaphroditus
Archytas of Tarentum (Architas Tarentinus) 132
Aristotle (Aristoteles) 18–20, 23, 24, 26, 69, 82, 129, 130
Aristoxenus of Tarentum (Aristoxenus Tarentinus) 70
Arnulf of Reims (Arnulphus Remensis) 114
Ascelin of Augsburg (Ascelinus Teutonicus) 87, 107
Asilo of Würzburg (Asilo Wirzburgensis) 53–55
Augustine (Aurelius Augustinus) 15, 18, 23, 27, 28, 78–80, 82, 186

B
Bede, the Venerable (Beda Venerabilis) 84, 86, 94, 97, 98, 106, 111, 132, 160
Benedict VIII, Pope (Theophylact of Tusculum, Benedictus VIII) 11, 122
Bernard of Clairvaux (Bernardus Claraevallensis) 185
Bernelin of Paris (Bernelinus iunior Parisius) 131–137, 141, 147, 152–154, 158, 162, 164–171, 180
Boethius (Anicius Manlius Torquatus Severinus) 15–31, 33–46, 48–53, 55–57, 65, 70–73, 78, 80–82, 85, 103, 114, 132, 165, 187
Byrhtfert of Ramsey (Bridfertus Ramesiensis) 160

C
Calcidius (Chalcidius) 42, 58, 85
Cassiodore (Flavius Magnus Aurelius Cassiodorus) 18–20, 23–26, 37, 38, 59, 78, 85, 88, 98, 110, 122
Censorinus (Grammaticus) 78
Charlemagne, Emperor (Carolus I Magnus) 10, 115, 122, 151
Charles II, the Bald, Emperor (Carolus II Calvus) 10
Ciacono, Alphons (Chacón, Alphonsus Ciacconius) 121
Cicero (Marcus Tullius) 82, 85, 185, 186
Constantine of Fleury and Micy (Constantinus Floriacensis / Miciacensis) 13, 33–35, 69–72, 75, 76, 78, 81, 92–95, 102, 114, 132, 156, 159, 165, 185, 186

D
Diogenes Laërtios (Diogenes Laertius) 186

E
Epaphroditus Agrimensor (Aprofiditus) 64
Eratosthenes of Cyrene (Erathostenes Cyrenaeus) 86, 110
Eriugena, John Scotus *see* John Scotus Eriugena
Euclid of Alexandria (Euclides Alexandriensis) 24, 25, 27–30, 56, 131
Eudoxus of Cnidus (Eudoxus Cnidius) 26
Evrard of Tours (Everardus Turonensis) 185, 186

F
Firmicus Maternus (Julius) 84
Fulbert of Chartres (Fulbertus Carnotensis) 12

G
Garin, abbot of Cuxa (Guarnerius) 157
Garland the Computist (Garlandus Compotista) 133
Geminus of Rhodes (Geminus Rhodius) 95
Gerald of Aurillac (Gerardus Auriliacensis) 157
Gerbert of Aurillac / of Reims / of Bobbio / of Ravenna, Pope Sylvester II (Gerbertus Auriliacensis / Remensis / Bobiensis / Ravennatensis, Silvester II): 12–14, 17, 33–36, 44, 48–51, 53, 55, 56, 58–68, 70–78, 80–97, 99–123, 126, 127, 129, 131–134, 136, 138, 142, 145, 148–151, 156–159, 162–165, 167, 170–183, 185–187
Gregory V, Pope (Bruno of Carinthia, Gregorius V) 11, 122

H
Harun al-Rashid (Aaron Rascidus) 115, 122
Henry II, Saint, Emperor (Henricus II Sanctus) 11, 60, 122
Heriger of Lobbes (Herigerus Lobiensis) 60, 133, 136, 158, 162–165, 167
Hermann the Lame / of Reichenau (Hermannus Contractus / Augiensis) 12, 53–55, 87, 107, 133
Horace (Quintus Horatius Flaccus) 82
Hugh Capet, king (Hugo Capetius) 114
Hyginus (Gaius Julius) 84, 87, 95

I
Iamblichus of Chalcis (Iamblichus Chalcidensis) 24, 26
Isidore of Sevilla (Isidorus Hispalensis) 15, 18, 20, 23, 24, 28–30, 37, 38, 52, 58, 59, 80, 85, 86, 88, 94, 97–99, 106, 110, 186

J
John XII, Pope (Octavianus Tusculanensis, Ioannes XII) 11
John XIII, Pope (John Crescentius, Ioannes XIII) 11, 70
John Scotus Eriugena (Iohannes Scotus Eriugena) 84, 86, 111, 112
John Stobaeus (Ioannes Stobaeus) 29
Joseph of Spain / the Wise (Ioseph Ispano / Sapiens) 157

Juvenal (Decimus Iunius Iuvenalis) 82

L

Laurent of Amalfi (Laurentius Amalfitanus) 133
Louis I, the Pious, Emperor (Ludovicus I Pius) 10
Lucan (Marcus Annaeus Lucanus) 82
Lupitus of Barcelona (Llobet, Sunifred / Lupitus Barchinensis) 83

M

Macrobius (Ambrosius Theodosius) 42, 85, 86, 88, 101, 109, 110, 112
Manilius (Marcus) 85
Marchetto of Padua (Marchetus Paduanus) 80
Marius Victorinus (Gaius, Victorinus Afer) 82
Martianus Capella (Minneus Felix) 18, 22–24, 29, 38, 58, 59, 78, 79, 84–87, 89, 95, 97–99, 101, 104, 106, 108, 111, 130, 131, 160
Miró Bonfill of Besalú / of Girona (Gerundensis) 157
Moderatus of Gades (Moderatus Gaditanus) 29

N

Nicomachus of Gerasa (Nicomachus Gerasensis) 16, 17, 19–22, 24–30, 37, 38, 41, 43, 55–57
Notker of Liège (Notgerus Leodiensis) 11, 33, 36, 37, 44–50, 53, 60

O

Odo of Cluny (Odo Cluniacensis) 133
Ohtric of Magdeburg (Ohtricus Magdeburgensis) 186
Oldoini, Augustine (Augustinus Oldoinus) 121
Otto I, the Great, Emperor (Otto I Magnus) 11, 70
Otto II, Emperor 11, 148–150, 186
Otto III, Emperor 11, 114, 122, 148, 156

P

Pacificus of Verona (Pacificus Veronensis) 123–127
Panatios of Rhodes (Panatius Rhodius) 185
Persius (Aulus Flaccus) 82
Plato 15, 23, 42, 51, 56, 58, 85, 129, 132
Pliny the Elder (Gaius Plinius Secundus) 85, 101, 111
Porphyry of Tyre (Porphyrius Tyrensis) 18, 82
[Pseudo-]Boethius *see* Anonymous, author called [Pseudo-]Boethius
Ptolemy (Claudius Ptolemaeus) 22, 84, 110
Pythagoras of Samos (Pythagoras Samius) 51, 53, 131, 132

R

Ralph of Laon (Radulphus Laudunensis) 132, 133
Raymond of Aurillac (Raimundus Auriliacensis) 185
Remigius of Auxerre (Remigius Autissiodorensis) 84, 131
Remigius of Trier (Remigius Treverensis) 91
Richer of Reims (Richerus Remensis) 34, 35, 51, 70, 75, 81–83, 85–88, 90–93, 97, 99–103, 115, 120, 126, 131, 134, 136, 147, 151, 156, 186, 187

Robert II, the Pious, king (Robertus II Pius) 114

S
Socrates 15
Statius (Publius Papinius) 82
Stobaeus, John *see* John Stobaeus

T
Terentius (Publius Varro) 82
Thales of Miletus (Thales Milesius) 24
Theophano, Empress (Theophanu imperatrix) 148
Theon of Smyrna (Theon Smyrnaeus) 17
Thietmar of Merseburg (Thiethmarus Merseburgensis) 93, 113, 116, 122, 123, 126, 127

Turchill the Computist (Turchillus Compotista) 130, 132, 133

V
Victorius of Aquitaine (Victorius Aquitanus) 13, 17, 36, 161, 164, 165, 167
Vigila of Albelda (Vigila Albeldensis) 151, 157, 163
Virgil (Publius Vergilius Maro) 82
Vitruvius Rufus Architect (Betrubus Rufus Architecton) 64

W
William of Malmesbury (Willelmus Malmesbiriensis) 35, 84, 113–115, 121, 126
Wyon, Arnold (Arnoldus Wionus) 121

Index of Personal Names (after 1700)

A
Albertson, D. 30
Ambrosetti, N. 130
Ampère, J.-J. 9

B
Bai, J. 78
Baillaud, B. 80
Bakhouche, B. 134
Barker, A. 70
Barnes, J. 16, 19
Baumann, H. 11
Beaujouan, G. 59, 117
Begbie, J. S. 78
Behrends, O. 64
Benson, R. L. 10
Bergmann, W. 64, 117, 132, 133, 134, 148, 156
Berschin, W. 12
Blandzi, S. 14
Boncompagni, B. 133
Borst, A. 12, 16, 53, 105, 117
Bower, C. M. 16, 75
Brockett, W. 148
Brown, N. M. 12, 158
Bubnov, N. 13, 35, 93, 156, 157, 172
Bud, R. 130
Burkert, W. 69
Burnett, C. 35, 130, 133, 136, 144, 148, 158
Butzer, P. L. 124

C
Caiazzo, I. 16, 36, 37
Caldwell, J. 19
Callebat, L. 59
Callu, J.-P. 35
Campogrossi Colognesi, L. 64
Cantor, M. 64
Carozzi, C. 114
Catalani, L. 10, 157
Cazeaux, M. 12
Celhoffer, M. 80
Černín, D. 14
Chadwick, H. 73, 76
Charbonnel, N. 76
Chasles, M. 172, 178
Cherniss, H. 69
Classen, A. 132
Constable, G. 11
Contreni, J. 9
Cornelli, G. 69
Corti, L. 20
Courtenay, L. T. 64
Craemer-Ruegenberg, I. 18
Crialesi, C. V. 16
Cristante, L. 79
Crossley, J. N. 30
Curd, P. 69

D
D'Onofrio, G. 10
Dachowski, E. 12, 70, 114
Darlington, O. G. 12, 34, 82
Davids, A. 10
De Gramont, J. 80
De Koninck, T. 18
De Rijk, L. M. 19
Dekker, E. 84
Delville, J.-P. 11
DeMayo, C. 34, 114
Desbordes, O. 59
Deza, E. 56
Deza, M. M. 56
Dohrn-van Rossum, G. 115

E
Eastwood, B. S. 101
Engelen, E.-M. 12
Erlande-Brandenburg, A. 12
Evans, G. R. 13, 16, 21, 36, 37, 52, 122, 146, 158

F
Fear, A. 80
Ferguson, W. K. 9
Flusche, A. M. 12, 70, 81, 83
Folkerts, M. 16, 53, 64, 65, 132, 133, 134, 138, 140, 142, 143, 148, 156, 157
Fournier, M. 22
Franci, R. 11, 17, 132
Frassetto, M. 10
Friedlein, G. 172, 176, 178
Frova, C. 13
Frost, G. 14

G
Gadducci, F. 10
Gagne, A. 18
Genin, C. 12
Germann, N. 105
Gibson, M. 16
Gibson, S. 70
Giovannetti, L. 16
Glenn, J. 103, 114
Goold, G. P. 85
Graham, D. W. 69
Guillaumin, J.-Y. 18, 22, 25, 26, 30, 59
Gurd, S. A. 70
Guthrie, S. R. 78
Guyotjeannin, O. 116

H
Hanke Jarošová, S. 14
Harrison, C. 78, 79
Haskins, Ch. H. 10, 11

Head, T. 35
Heinzer, F. 12
Hellmann, M. 12
Hen, Y. 123
Henderson, J. 80
Herlinger, J. W. 80
Herren, M. 10
Hexter, R. 20
Hiatt, A. 110
Hicks, A. 19, 20
Hill, D. R. 115
Holmes, U. T. 9, 11
Honigman, E. 110
Hue, D. 80
Huffman, C. A. 69, 70
Hughes, B. 16
Huglo, M. 75

I
Innes, M. 123
Iung, J. E. 76

J
Jelínek, I. 14
Junius, H. 64
Juste, D. 84

K
Karp, A. 25
Karpati, A. 16
Katz, V. J. 16
Kaylor, N. H. Jr. 16
Kibre, P. 18
Klinkenberg, H. M. 18
Knobloch, E. 64
Kupper, J.-L. 11
Kurth, G. 11

L
La Croix, R. R. 78
La Rocca, C. 123
Laffineur-Crepin, M. 11

Lake, J. 82, 103, 186
Lattin, Pratt H. 35, 60, 157
Landes, R. 10
Le Goff, J. 9
Leach, E. E. 81
Leonardi, C. 9
Lichfield, M. 70
Light, L. 103
Lindberg, D. C. 18
Lindgren, U. 34, 70, 85, 87, 122, 134
Lohrmann, D. 124
Lutz, C. E. 12, 35

M
MacKinney, L. C. 12
Marenbon, J. 9, 19
Masi, M. 18, 21, 27, 30, 71
Materni, M. 59
Mathiesen, T. J. 78, 79, 81
Matthews, M. R. 115
Mazur, J. 132
McCluskey, S. C. 82, 113
McCready, W. D. 110
McKitterick, R. 9, 10
Melve, L. 11
Meyer, C. 76
Meyer, H. 27
Michel, H. 124
Miller, G. A. 61, 64
Moller, C. 64
Moretti, G. 79
Mostert, M. 12
Moyer, A. E. 16, 53, 81
Mulke, M. 9

N
Naumann, H. 10
Nelson, J. L. 9, 11
North, J. D. 13
Novikoff, A. J. 11
Nuvolone, F. G. 75, 76, 133, 148, 151

O
O'Connor, R. 9, 10
Obbink, D. 69
Obertello, L. 18, 21
Oestmann, G. 126
Oldoni, M. 12, 185
Olleris, A. 172

P
Pagli, P. 17, 132
Panvini-Carciotto, M. G. 35
Patzelt, E. 9
Peden, A. M. 13, 36, 122, 158
Petrovićová, K. 79
Petrucci, F. M. 17
Phillips, P. E. 16, 18, 30
Pizzani, U. 21, 71
Polara, G. 9
Poleg, E. 103
Poulle, E. 116
Psík, R. 14
Pullan, J. M. 130

R
Radke, G. 28
Reilly, D. J. 103
Ribemont, B. 80
Riche, P. 9, 12, 13
Rigatelli, L. T. 17, 132
Rimple, M. T. 30, 75
Roche, J. J. 13
Rollo, D. 115
Roskam, G. 185
Rossi, P. 35, 60
Rouche, M. 12
Rowett, C. 69
Rubenstein, J. 10
Ruckert, O. 10
Ruggles, C. L. N. 82
Ryan, W. F. 130
Rywiková, D. 14

S
Sachs, K. J. 51, 75
Saenger, P. 157
Saliba, G. 85
Samso, J. 35
Savage-Smith, E. 90, 91, 95
Scharling, A. 34, 157
Schmid, H. 60
Schrade, L. 80
Schubring, G. 25
Schulmeyer-Ahl, K. 11
Selin, H. 115
Shelby, L. R. 58, 64
Sider, D. 69
Sigismondi, C. 14, 35, 53, 71, 75, 149
Sigismondi, F. 13
Silva, J. N. 53, 157
Šíma, A. 69
Smith, D. E. 130, 132
Speer, A. 18
Špelda, D. 14, 99
Stahl, W. H. 23, 27, 110
Stautz, B. 117
Stevens, W. M. 117
Studtmann, P. 19
Sugden, K. F. 130
Suntrup, R. 27
Swanson, R. N. 11

T
Tarrant, H. 17
Tavosanis, M. 10
Teeuwen, M. 23, 79
Thompson, R, M. 116
Thomson, R. B. 13
Thorndike, L. 115
Tosi, M. 12, 13, 35, 59, 114, 121, 122

Townsend, D. 20
Trompf, G. W. 9
Truitt, E. R. 10, 116
Turner, A. J. 117

U
Udina i Martorell, F. 35
Usher, A. P. 115

V
Van de Vyver, A. 13, 84, 95, 117
Vasina, A. 121
Vaughn, S. N. 10
Verger, J. 9
Vogel, C. 16
Von Gotstedter, A. 117

W
Wagner, D. L. 27
Walden, D. K. 21
Wallis, F. 80
Wangerin, L. E. 11
Warner, D. J. 130
Warren, F. M. 35
Wedell, M. 132
Weisheipl, J. A. 18
Wessely, O. 60
Wiesenbach, J. 124
Williams, J. R. 10, 34
Williams, P. 71, 81
Winterbottom, M. 116
Wood, I. 9
Wood, J. 80

Z
Zimmermann, H. 122
Zotz, T. 12
Zuccato, M. 35, 85, 91

**Philosophy and Cultural Studies Revisited /
Historisch-genetische Studien zur Philosophie und Kulturgeschichte**

Editd by Seweryn Blandzi

Vol.	1	Dorota Muszytowska / Janusz Kręcidło / Anna Szczepan-Wojnarska (eds.): Jerusalem as the Text of Culture. 2018.
Vol.	2	Dorota Probucka (ed.): Contemporary Moral Dilemmas. 2018.
Vol.	3	Ján Zozuľak: Inquiries into Byzantine Philosophy. 2018.
Vol.	4	Marek Piechowiak: Plato's Conception of Justice and the Question of Human Dignity. 2018.
Vol.	5	Dorota Probucka: Ethics in Ancient Greece and Rome. 2019.
Vol.	6	Imelda Chłodna-Błach: From Paideia to High Culture. The Philosophical and Anthropological Foundation of the Dispute about Culture. 2020.
Vol.	7	Arkadiusz Jabłoński / Jan Szymczyk: REALIST-AXIOLOGICAL PERSPECTIVES AND IMAGES OF SOCIAL LIFE. A Century of Sociology at the John Paul II Catholic University of Lublin. 2020.
Vol.	8	Marek Piechowiak: Plato's Conception of Justice and the Question of Human-Dignity. Second Edition, Revised and Extended. 2021.
Vol.	9	Ewa Szumilewicz: On the Paradox of Cognition. 2021.
Vol.	10	Marek Otisk: Arithmetic in the Thought of Gerbert of Aurillac. 2022

www.peterlang.com

www.ingramcontent.com/pod-product-compliance
Ingram Content Group UK Ltd.
Pitfield, Milton Keynes, MK11 3LW, UK
UKHW041923210426
5322IPUK00002B/28